Safety Assessment for Chemical Processes

Safety Assessment for Chemical Processes

Contributors

Kang Sun, Long Bai et al.

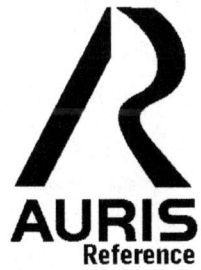

AURIS
Reference

www.aurisreference.com

Safety Assessment for Chemical Processes

Contributors: Kang Sun, Long Bai et al.

Published by Auris Reference Limited

www.aurisreference.com

United Kingdom

Safety Assessment for Chemical Processes

ISBN: 978-1-78154-894-3

British Library Cataloguing in Publication Data
A CIP record for this book is available from the British Library

Printed in the United Kingdom

Exclusively distributed by CBS Publishers & Distributors Pvt. Ltd.

Sales & Distribution Rights only for India, Pakistan, Bangladesh, Sri Lanka, Nepal and Bhutan. This book is not to be sold outside these territories.

Contents

List of Abbreviations

AAC	Azide Alkyne Cycloaddition
ANN	Artificial Neural Network
ANOVA	Analysis of Variance
CCA	Cause-Consequence Analysis
CCRD	Central Composite Rotational Design
EFC	Ethanol Fueled Fuel Cell
ESR	Ethanol Steam Reforming
ETA	Event Tree Analyses
FCI	Fixed capital investment
FMEA	Failure Modes and Effects Analysis
FTA	Fault Tree Analysis
GHG	Greenhouse Gas
HFS	High Fructose Syrups
HMDB	Human Metabolome database
HPG	Hybrid Power Generation
HRA	Human Reliability Analysis
ICA	Independent Component Analysis
ID	Identity Number
ITA	Isoplethic Thermal Analysis
KPCA	Kernel principal component analysis
LOPA	Layer of protection analysis
LTWGS	Low Temperature Water Gas-Shift
MM	Molecular Mass
MNP	Magnetic Nanoparticles
MOI	Multiplicity of Infection
MW	Microwave
NPV	Net Present Value
OSHA	Occupational Safety and Health Administration
PEG	Polyethylene Glycol
PET	Poly-Ethylene Terephthalate
PFD	Process Flow Diagram
PLA	Poly-Lactic Acid
PSA	Pressure Swing Adsorption
PV	Photovoltaic
PWM	Pulse-Width Modulation
RBC	Red Blood Cell
SAR	Structure-Activity Relationships
SAWS	State Administration of Work Safety)
SPE	Squared Prediction Error
STITCH	Search Tool for Interactions of Chemicals
TAC	Total Annualized Cost

VECM	Vector Error Correction Model
WBC	White Blood Cell
WNN	Wavelet Neural Network

List of Contributors

Kang Sun
Center for Studies of Marine Economy and Sustainable Development, Liaoning Normal University, Dalian, China

Long Bai
Center for Studies of Marine Economy and Sustainable Development, Liaoning Normal University, Dalian, China

Xiaohong Li
Center for Studies of Marine Economy and Sustainable Development, Liaoning Normal University, Dalian, China

Hongbo Wang
Key Laboratory of Molecular Pharmacology and Drug Evaluation (Ministry of Education of China), School of Pharmacy, Yantai University, Yantai, China

Jianqiao Zhang
Key Laboratory of Molecular Pharmacology and Drug Evaluation (Ministry of Education of China), School of Pharmacy, Yantai University, Yantai, China

Guangyao Lv
Key Laboratory of Molecular Pharmacology and Drug Evaluation (Ministry of Education of China), School of Pharmacy, Yantai University, Yantai, China

Jinbo Ma
Department of clinical medicine, Binzhou Medical College, Yantai, China

Pengkai Ma
Key Laboratory of Molecular Pharmacology and Drug Evaluation (Ministry of Education of China), School of Pharmacy, Yantai University, Yantai, China

Guangying Du
Key Laboratory of Molecular Pharmacology and Drug Evaluation (Ministry of Education of China), School of Pharmacy, Yantai University, Yantai, China

Zongliang Wang
Key Laboratory of Molecular Pharmacology and Drug Evaluation (Ministry of Education of China), School of Pharmacy, Yantai University, Yantai, China

Jingwei Tian
Key Laboratory of Molecular Pharmacology and Drug Evaluation (Ministry of Education of China), School of Pharmacy, Yantai University, Yantai, China

State Key Laboratory of Long-acting and Targeting Drug Delivery Technologies (Luye Pharma Group Ltd.), Yantai, China

Weishuo Fang
State Key Laboratory of Bioactive Substances and Functions of Natural Medicines, Institute of Materia Medica, Chinese Academy of Medical Sciences and Peking Union Medical College, Beijing, China

Fenghua Fu
Key Laboratory of Molecular Pharmacology and Drug Evaluation (Ministry of Education of China), School of Pharmacy, Yantai University, Yantai, China

Lynn El Haddad
De´partement de biochimie et de microbiologie, Faculte´ des sciences et de ge´nie, Groupe de recherche en e´cologie buccale, Faculte´ de me´decine dentaire, Fe´lix d'He´relle Reference Center for Bacterial Viruses, Universite´ Laval, Que´bec, Canada

Nour Ben Abdallah
Food Research and Development Centre, Agriculture and Agri-Food Canada, SaintHyacinthe, Que´bec, Canada

Pier-Luc Plante
De´partement de Me´decine Mole´culaire, Faculte´ de Me´decine, Universite´ Laval, Que´bec, Canada

Jeannot Dumaresq
De´partement de Microbiologie et d'Infectiologie, Centre Hospitalier Affilie´ Universitaire Hoˆtel-Dieu de Le´vis, Le´vis, Que´bec, Canada

Ramaz Katsarava
Institute of Chemistry & Molecular Engineering, Agricultural University of Georgia, University Campus at Digomi, Tbilsi, Georgia

Steve Labrie
De´partement des sciences des aliments et de nutrition, Faculte´ des sciences de l'agriculture et de l'alimentation, Dairy Science and Technology Research Centre/Institute of nutrition and functional foods, Universite´ Laval, Que´bec, Canada

Jacques Corbeil
De´partement de Me´decine Mole´culaire, Faculte´ de Me´decine, Universite´ Laval, Que´bec, Canada

Daniel St-Gelais
Food Research and Development Centre, Agriculture and Agri-Food Canada, SaintHyacinthe, Que´bec, Canada

De´partement des sciences des aliments et de nutrition, Faculte´ des sciences de l'agriculture et de l'alimentation, Dairy Science and Technology Research Centre/Institute of nutrition and functional foods, Universite´ Laval, Que´bec, Canada

Sylvain Moineau
De´partement de biochimie et de microbiologie, Faculte´ des sciences et de ge´nie, Groupe de recherche en e´cologie buccale, Faculte´ de me´decine dentaire, Fe´lix d'He´relle Reference Center for Bacterial Viruses, Universite´ Laval, Que´bec, Canada

E. Labarthe
Laboratoire Hydrazines et Composés Energétiques Polyazotés, UMR CNRS-CNES-SME-Groupe Safran, Université Claude Bernard Lyon 1, Villeurbanne, France

A. J. Bougrine
Laboratoire Hydrazines et Composés Energétiques Polyazotés, UMR CNRS-CNES-SME-Groupe Safran, Université Claude Bernard Lyon 1, Villeurbanne, France

Véronique Pasquet
Laboratoire Hydrazines et Composés Energétiques Polyazotés, UMR CNRS-CNES-SME-Groupe Safran, Université Claude Bernard Lyon 1, Villeurbanne, France

H. Delalu
Laboratoire Hydrazines et Composés Energétiques Polyazotés, UMR CNRS-CNES-SME-Groupe Safran, Université Claude Bernard Lyon 1, Villeurbanne, France

Juliano Missau
Department of Chemical Engineering, Federal University of Santa Maria, Santa Maria 97105-900, Brazil

Amir J Scheid
Department of Chemical Engineering, Federal University of Santa Maria, Santa Maria 97105-900, Brazil

Edson L Foletto
Department of Chemical Engineering, Federal University of Santa Maria, Santa Maria 97105-900, Brazil

Sergio L Jahn
Department of Chemical Engineering, Federal University of Santa Maria, Santa Maria 97105-900, Brazil

Marcio A Mazutti
Department of Chemical Engineering, Federal University of Santa Maria, Santa Maria 97105-900, Brazil

Raquel C Kuhn
Department of Chemical Engineering, Federal University of Santa Maria, Santa Maria 97105-900, Brazil

Brian S. Flowers
Department of Chemical and Biological Engineering, The University of Alabama, Box 870203, Tuscaloosa, AL 35487-0203, USA

Ryan L. Hartman
Department of Chemical and Biological Engineering, The University of Alabama, Box 870203, Tuscaloosa, AL 35487-0203, USA

Rajender S Varma
Sustainable Technology Division, National Risk Management Research Laboratory, U.S. Environmental Protection Agency, 26 West Martin Luther King Drive, MS 443, Cincinnati, Ohio 45268, USA

Susmit S Bapat
School of Chemical Engineering, Oklahoma State University, Stillwater, OK 74078, USA

Clint P Aichele
School of Chemical Engineering, Oklahoma State University, Stillwater, OK 74078, USA

Karen A High
School of Chemical Engineering, Oklahoma State University, Stillwater, OK 74078, USA

Wei Wu
Department of Chemical Engineering, National Cheng Kung University, Tainan 70101, Taiwan

Yuan-Tai Hsu
National Yunlin University of Science and Technology, Douliou, Yunlin 64002, Taiwan.

Po Chih Kuo
Department of Chemical Engineering, National Cheng Kung University, Tainan 70101, Taiwan

Ramzan Naveed
Department of Chemical Engineering, University of Engineering and Technology, Lahore, Pakistan

Zeeshan Nawaz
Chemical Technology Development, STCR, Saudi Basic Industries Corporation (SABIC), Kingdom of Saudi Arabia

Werner Witt
Lehrstuhl Anlagen und Sicherheitstechnik, Brandenburgicshe Technische Universität, Cottbus, Germany

Shahid Naveed
Department of Chemical Engineering, University of Engineering and Technology, Lahore, Pakistan

Lei Chen
College of Information Engineering, Shanghai Maritime University, Shanghai, China

Jing Lu
Drug Discovery and Design Center (DDDC), Shanghai Institute of Materia Medica, Shanghai, China

Jian Zhang
Department of Ophthalmology, Shanghai First People's Hospital Affiliated to Shanghai Jiaotong University, Shanghai, China

Kai-Rui Feng
Simcyp Limited, Blades Enterprise Centre, Sheffield, United Kingdom

Ming-Yue Zheng
Drug Discovery and Design Center (DDDC), Shanghai Institute of Materia Medica, Shanghai, China

Yu-Dong Cai
Institute of Systems Biology, Shanghai University, Shanghai, China

Lijie Guo
Hebei Key Laboratory of Applied Chemistry, College of Environmental and Chemical Engineering, Yanshan University, Qinhuangdao, Hebei 066004, China

Jianxin Kang
Hebei Key Laboratory of Applied Chemistry, College of Environmental and Chemical Engineering, Yanshan University, Qinhuangdao, Hebei 066004, China

Suhendra Werner
Institute of Plant Design and Safety Technology, Technical University of Brandenburg LAS-BTU Cottbus, Haus 213, Burgerchausse 2-3, 03044-Cottbus, Germany
Departement of Chemical Engineering, Faculty of Industrial Technology, Ahmad Dahlan University, 31. Prof DR. Soepomo, Janturan, Yogyakarta, DIY-Indonesia

Witt Fred
Institute of Plant Design and Safety Technology, Technical University of Brandenburg
LAS-BTU Cottbus, Haus 213, Burgerchausse 2-3, 03044-Cottbus, Germany

Compart
Institute of Plant Design and Safety Technology, Technical University of Brandenburg
LAS-BTU Cottbus, Haus 213, Burgerchausse 2-3, 03044-Cottbus, Germany

Preface

A knowledge of at least the fundamentals of chemical safety technology is indispensable for chemists and engineers working in chemical industry. In spite of the good safety records of chemical plants many people regard chemical production as dangerous because of a few major accidents that have occurred. The investigation of risks, and preventive measures to be taken to minimize the probability of an accident, as well as its consequences are explained. Safety Assessment for Chemical Processes covers topics in the field of chemical safety and chemical industry. In first chapter, an empirical test on the investment performance of chemical safety facilities will be implemented to talk about the cause of the chemical accidents happening frequently through analyzing the dynamic relationship between the investment and the performance of chemical safety facilities, using the time series data by VECM. In second chapter, the pharmacokinetics, biodistribution, antitumor efficacy and safety characteristics of liposome-based Lx2-32c were explored and compared with those of cremophor-based Lx2-32c. Third chapter presents a study suggests that some staphylococcal phages can be propagated on food-grade bacteria for biocontrol and safety purposes. Fourth chapter presents a new strategy for the synthesis of N-aminopiperidine (NAPP). Fifth chapter focuses on the immobilization of commercial inulinase on alginate. Sixth chapter presents a framework on particle handling techniques in microchemical processes. In seventh chapter the uses of magnetically recyclable nano-catalysts for a variety of organic reactions are described in conjunction with activation via microwave irradiation. Eighth chapter focuses on the development of a sustainable process for the production of polymer grade lactic acid. Ninth chapter proposes an EFC/PV/Battery based hybrid power generation system to meet 24-hour power demand. In tenth chapter a systematic methodology based on independent modules and its different stages to deal this problem is presented in detail. In eleventh chapter, an order-classifier was built to predict a series of toxic effects based on data concerning chemical-chemical interactions under the assumption that interactive compounds are more likely to share similar toxicity profiles. Twelfth chapter focuses on the approach of a hybrid process monitoring and fault diagnosis approach for chemical plants. Last chapter reviews some most frequently causes of distillation column malfunction. First, analysis of case histories will be discussed for providing guidelines in identifying potential trouble spots in distillation column. A dynamic simulation for operational failure is simulated as the basis for assessing the consequences.

Chapter 1

ANALYSIS OF THE CHEMICAL SAFETY FACILITY INVESTMENT PERFORMANCE IN CHINA

Kang Sun, Long Bai, Xiaohong Li

Center for Studies of Marine Economy and Sustainable Development, Liaoning Normal University, Dalian, China

ABSTRACT

This paper adopts the accident incidence, the gross industry output value, the investment in safety facilities, and per capita wage of employment as the indexes to empirically analyze the investment performance of chemical safety facilities using time series data by VECM in China. The empirical results indicate that for China's chemical industry, increasing investment fails to improve the short-term safety level significantly because of the offsetting behavior of workers. Over the long term, the offsetting behavior tends to disappear, and the chemical accident incidence can be decreased through increasing investment. Poor safety awareness among workers is one of the causes of accident incidences. The conclusions provide theoretical support for China to perfect chemical industry safety management.

INTRODUCTION

With the birth of chemical industry, controversy arose because of the industry's high risk to human safety and to the environment. In recent decades, hazardous chemical accidents have become a worldwide problem. Many international conventions relating to the chemical industry have been legislated, including regulations, policies, and industry management systems, such as the Rotterdam Convention (1998), the Stockholm Convention (2001), the UNEP (United Nations Environment Programme) Strategic Approach to International Chemicals Management (1998), and the EU (European Union) directive on Registration, Evaluation, Authorization and Restriction of Chemicals (REACH, 2007). Over the past 20 years, China has become a major international player in the chemical industry. Because of the rapid development of China's chemical industry, it has become the pillar industry of the national economy. Consequently, hazardous chemical accidents happen frequently, often engendering secondary disasters. Secondary disasters also pose big threats to human safety and health. The safety, health, and environmental problems caused by hazardous chemical accidents are increasing. The high frequency of chemical accidents has caused serious social problems in China. China has recently enacted more than 200 laws, administrative regulations, and departmental rules for its chemical industry. This intensification of management is unprecedented, but it still cannot stop the rise of chemical accidents year by year [1] . Thus, it is necessary to analyze the safety management effect empirically for China's chemical industry.

Most scholars used econometric analysis methods to research how the laws and regulations strengthened by safety regulatory organization influence the number of safety accidents brought by production and safety regulation effects empirically. Smith (1979) suggested that OSHA (Occupational Safety and Health Administration) policies were effective after analyzing the impact of its inspection on manufacturing injury rates over the period of 1973-1974 [2] . Gray and Scholz (1993) analyzed the industry panel data of 1979-1985 and concluded that OSHA policies reduced workplace fatalities by 22% [3] . In addition, Beck and Alford [4] (1980), Carmichael [5] (1986), and Weil [6] (1996) also deemed safety regulation was effective. Nevertheless, the result of Viscusi's [7] [8] (1979, 1992) study was that the more the regulation was promulgated, the more deaths were caused through researching the U.S. government's safety and health regulation policy. Other scholars [9] [10] found that the regulation effect was unsatisfactory due to the regulatory capture phenomenon. The extant literature about chemical safety mainly focuses on accident disposal [11] , accident statistical analysis [12] , and the domino effect [13] . The literature in relation to the cause of chemical accidents which happened frequently is relatively lacking, especially empirical analysis in China .

In this paper, an empirical test on the investment performance of chemical safety facilities will be implemented to talk about the cause of the chemical accidents happening frequently through analyzing the dynamic relationship between the investment and the performance of chemical safety facilities, using the time series data by VECM (Vector Error Correction Model).

SELECTION CRITERIA AND DATA

The general empirical analysis approach for regulation effect is to do a regression of the regulation behavior index to the regulation effect index, then test the significance and direction of the influence. Due to the particularity and complexity of the chemical industry, there are many factors that influence the chemical accident incidence. In this paper, index selection is mainly on the basis of existing literature. For instance, safety inspections and fines of factories had been selected as regulation behavior; workplace accident mortality had been selected as regulation effect by Klick and Stratmann (2003) in their research using the data provided by the OSHA [14] . Similarly, in this paper, in order to highlight the chemical accidents which occurred frequently in China , the accident incidence is selected as investment performance index. In order to avoid the contingency factors of accident incidence increases, we normalize the index by 100 million Yuan output value. As for the investment of chemical safety facilities, we choose the investment in fixed assets to denote investment in chemical safety facilities[1]. Following the Peltzman effect[2], the workers' offsetting behavior is added to the empirical analysis. According to the rational economic man principle, per capita wage is selected to signify the workers' offsetting behavior. In addition to chemical enterprise behavior and worker behavior, other disturbance factors also influence the performance. In order to control these influences, the chemical gross industry output value index also is included in the empirical model.

This paper uses the annual data[3] over the period 1981-2011 to identify the development trend of China's chemical accidents. Chemical industry refers to the Manufacture of Raw Chemical Materials and Chemical Products according to national economy classifications in China . The number of chemical accidents (1981-2000) is drawn from the book Selected Cases of Major Chemical Accidents, and the number of chemical accidents (2001-2011) is from the AIS (Accident Inquiry System) of the SAWS (State Administration of Work Safety) in China. The data of chemical gross industry output value and investment in fixed assets are calculated from the China Statistical Yearbook (1982-2012). The data about per capita wage is calculated from the China Labor Statistical Yearbook (1991-2012). In order to eliminate price change effects on chemical gross industry output value and investment in fixed assets, we transform the data of chemical gross industry output value and investment in fixed assets to constant price of 1978 by the GDP deflator. To eliminate heteroscedasticity, except for accident incidence, all of the other variables take the form of natural logarithms. These four variables are respectively expressed by S_t, V_t, I_t, W_t.

EMPIRICAL RESULTS

Unit Root Test

We begin our empirical analysis by testing for unit roots in the accident incidence S_t, gross industry output value V_t, investment in safety facilities I_t, and per capita wage W_t, because the integrational properties are crucial for the cointegration test and Granger causality test in VECM framework. We apply the conventional augmented Dickey and Fuller (ADF, 1979) test to establish the integrational properties of S_t, V_t, I_t, W_t. The calculated t-statistics together with the lag length selected using the SIC (Schwarz Information Criterion), as well as the critical value at 5% for the accident incidence S_t, gross industry output value V_t, investment in safety facilities I_t, and per capita wage W_t series are reported in Table 1.

In Table 1, the calculated t-statistics for the levels of accident incidence S_t, gross industry output value V_t, investment in safety facilities I_t, and per capita wage W_t series are greater than the critical value at the 5% significant level. This implies that we cannot reject the unit root null hypothesis. However, when we convert the accident incidence S_t, gross industry output value V_t, investment in safety facilities I_t, and per capita wage W_t series into first difference and subject the series to the ADF test, the calculated t-statistic for all accident incidence S_t, gross industry output value V_t, investment in safety facilities I_t, and per capita wage W_t is smaller than the critical value at the 5% level. This implies that we can reject the unit root null hypothesis for all series in first difference form. As a result, all variables are integrated of order one. This paves the way for conducting tests for cointegration and Granger causality in a VECM framework later in the paper.

Cointegration Test

In order to examine the long-term relationship among the accident incidence S_t, gross industry output value V_t, investment in safety facilities I_t, and per capita wage W_t, we use the Johansen test method to perform a cointegration test. According to the AIC (Akaike Information Criterion) and SIC (Schwarz Information Criterion), the optimal number of lags for the Johansen cointegration test method is 3. When the trace statistic is greater than the critical value at the 5% significant level, reject the null hypothesis of "no cointegration"; When the trace statistic is smaller than the critical value at the 5% significant level, accept the null hypothesis. The test results are listed in Table 2.

As illustrated in Table 2, we accept the null hypothesis of "at most 3 cointegration relationships existed" among the accident incidence S_t, gross output value V_t, investment in safety facilities I_t, and per capita wage W_t at the 5% significant level. The cointegration equation (two additional cointegration relationships are omitted because they are irrelevant to this paper) is estimated below (the standard errors are in parentheses):

$$EC = S - 1.219691V + 0.283725I + 1.435979W$$
$$\quad\quad\quad (0.12454) \quad\quad (0.08206) \quad\quad (0.13543)$$

(1)

Table 1. ADF unit root test results of S_t, V_t, I_t, W_t.

Variable	S_t	ΔS_t	V_t	ΔV_t	I_t	ΔI_t	W_t	ΔW_t
t-statistic	−0.9979	−5.3515	−1.0925	−5.3710	−1.7326	−2.3890	13.0112	−4.3308
[LL]	[0]	[0]	[0]	[0]	[1]	[0]	[0]	[0]
CV	−1.9525	−1.9529	−3.5684	−3.5742	−3.5742	−1.9529	−1.9525	−3.5742

Notes: LL denotes lag length, which is selected using the SIC automatically and CV denotes critical values at the 5% significant level. D is the difference operator.

Table 2. Cointegration test results of S_t, V_t, I_t, W_t

Hypothesized No. of CE(s)	Eigenvalue	Trace Statistic	0.05 Critical Value	P
None*	0.9323	146.8033	47.8561	0.0000
At most 1*	0.7701	74.1057	29.7971	0.0000
At most 2*	0.7140	34.4119	15.4947	0.0000
At most 3	0.0224	0.6104	3.8415	0.4346

From Equation (1), in the long term, an increase in the gross output value V_t has a positive effect on accident incidence S_t, and if the gross output value V_t is increased by 1%, the accident incidence S_t will be increased by 1.22%. An increase of investment in safety facilities I_t and per capita wage W_t both have negative effects on accident incidence S_t, and when investment in safety facilities I_t and per capita wage W_t are respectively increased by 1%, the accident incidence S_t will be decreased separately by 0.28% and 1.44%. The negative effect of per capita wage W_t is stronger than the positive effect of gross output value V_t.

VECM

A causality test is often used to analyze the causal relationship among the variables. When there is cointegration relationship among the variables, we can construct a VECM to get the regression equation including the error correction item. The Wald joint test is then used to test the significance of the coefficient both on the variables and the error correction item to judge the causality direction [15] [16] in the VECM framework. The lag length is equal to the lag length for the cointegration test. The general equation of VECM is expressed by:

$$\Delta Y_t = \sum_{i=1}^{p} \Gamma_i \Delta y_{t-i} + \lambda EC_{t-1} + \varepsilon_t$$

(2)

Where $Y_t = \begin{bmatrix} S_t & V_t & I_t & W_t \end{bmatrix}'$, Γ_i is the coefficient matrix, reflecting the impact made by short-term change

of explaining variables to short-term change of explained variable. EC_{t-1} is the error correction item, reflecting the long-term equilibrium relationship of variables. 1 is the coefficient vector of EC_{t-1}, reflecting the adjustment velocity from disequilibrium to equilibrium when it deviates from a long-term equilibrium state. ε_t denotes random error vector.

The specific VECM equation[4] in which accident incidence S_t is the explained variable with gross industry output value V_t, investment in safety facilities I_t and per capita wage W_t as the explaining variable is:

$$\Delta S_t = \begin{bmatrix} -0.1976 \\ -0.0158 \\ -0.0134 \\ 0.0712 \end{bmatrix} \begin{bmatrix} \Delta S & \Delta V & \Delta I & \Delta W \end{bmatrix}_{t-1} + \begin{bmatrix} -0.0510 \\ 0.0100 \\ 0.0062 \\ -0.0083 \end{bmatrix} \begin{bmatrix} \Delta S & \Delta V & \Delta I & \Delta W \end{bmatrix}_{t-2}$$

$$+ \begin{bmatrix} -0.2474 \\ -0.0279 \\ -0.0195 \\ 0.1598 \end{bmatrix} \begin{bmatrix} \Delta S & \Delta V & \Delta I & \Delta W \end{bmatrix}_{t-3} - 0.0146 EC_{t-1} - 0.0108.$$

(3)

Where $EC_{t-1} = S_{t-1} - 1.2197 V_{t-1} + 0.2837 I_{t-1} + 1.4360 W_{t-1} - 3.3581$.

From the Equation (3), the coefficient on EC_{t-1} is −0.0146, meaning that the adjustment degree of the disequilibrium for the previous year is the 1.46%.

At the 5% significant level, use the Wald joint test on the equation in which accident incidence S_t is the explained variable, with gross industry output value V_t, investment in safety facilities I_t and per capita wage W_t as explaining variables. The null hypothesis of the Wald joint test is that there is no granger causality between variables. When the probability value of the Wald joint test $\square^2$ is greater than 0.05, accept the null hypothesis; When the probability value of the Wald joint test $\square^2$ is smaller than 0.05, rejected the null hypothesis. The Wald joint test results are shown in Table 3.

The statistical significances of each coefficient in Table 3 indicate that at the 5% significant level, all the null hypothesizes (H_0) were rejected. Combined with the VECM equation interpretation, we know that in both the short and long term, the changes of gross industry output value V_t, investment in safety facilities I_t and per capita wage W_t are all the Granger causes of change in accident incidence S_t.

Impulse Response Function

Through impulse response function analysis, the path of influence affected by gross industry output value V_t, investment in safety facilities I_t and per capita wage W_t on current value and future value of accident incidence S_t can be obtained. To avoid variable order affecting the results, we choose the generalized impulse response function to do impulse response function analysis. The impulse response curves are represented in Figure 1 and Figure 2. The horizontal axis denotes the period (here we report only 10 periods; increasing the period does not affect the conclusion), and the vertical axis denotes the response degree.

Figure 1 shows the curve of response of per capita wage W_t to investment in safety facilities I_t impulse. From Figure 1 we can demonstrate that, in the short term, the response of per capita wage W_t to investment in safety facilities I_t impulse is positive. In the long term, the response is negative. For a positive information rush of I_t, the maximum positive response of W_t is achieved in the first period. Starting from the third period, the response turns from positive to negative. The strongest negative response is in the fourth stage, subsequently decreasing gradually. That is, in the short term, an increase in investment in safety facilities will increase the per capita wage of workers. In the long run, the increase investment in safety facilities can reduce per capita wage.

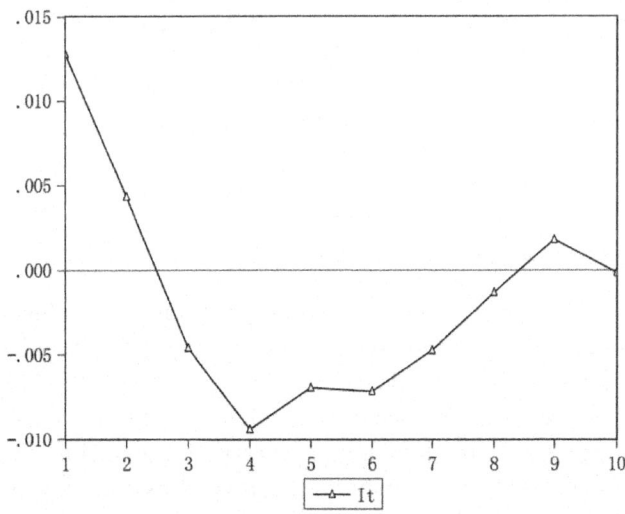

Figure 1. Curve of response of W_t to I_t.

Table 3. The results of the Granger causality test of S_t, V_t, I_t, W_t

		Test equation[5]: $\Delta S_t = \sum_i [\alpha_{1i}\Delta S_{t-i} + \beta_{1i}\Delta V_{t-i} + \gamma_{1i}\Delta I_{t-i} + \delta_{1i}\Delta W_{t-i}] + \lambda_1 EC_{t-1} + \varepsilon_t$				
		ΔV_t	ΔI_t	ΔW_t	Joint test	EC test
	H_0	$\beta_{1i} = 0$	$\gamma_{1i} = 0$	$\delta_{1i} = 0$	$\beta_{1i} = \gamma_{1i} = \delta_{1i} = 0$	$\lambda_1 = 0$
ΔS_t	χ^2	14.8771	8.3034	20.0818	42.7323	4.03778
	P	0.0019	0.0401	0.0002	0.0000	0.0445

NOTE: H_0 indicates that the row variable doesn't cause the column variable; P is the probability value of Wald joint test c^2.

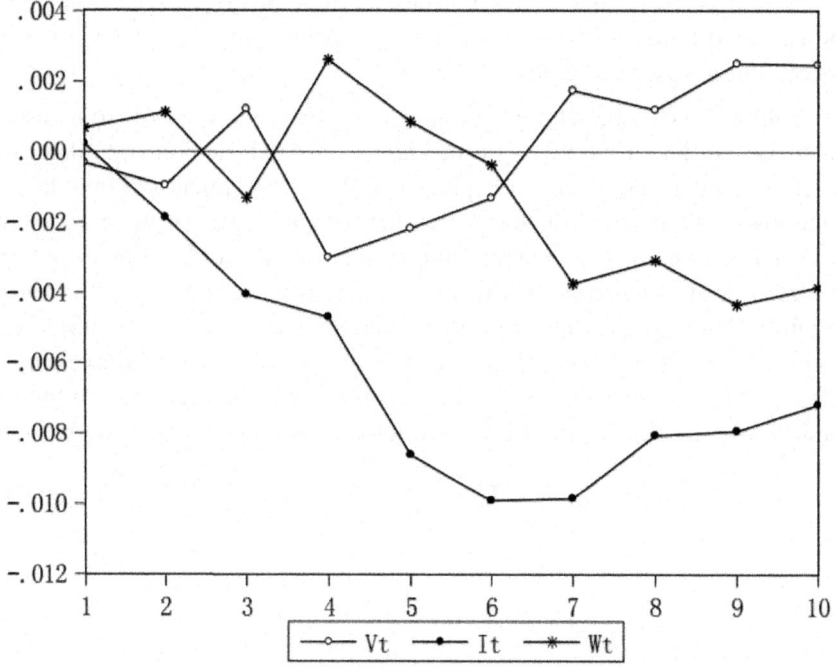

Figure 2. Curves of response of S_t to V_t, I_t, W_t.

Figure 2 shows the curves of response of accident incidence S_t to gross industry output value V_t impulse, investment in safety facilities I_t impulse, and per capita wage W_t impulse. From Figure 2 we can deduce: (1) in the short term, the response of accident incidence S_t to gross industry output value V_t is negative. In the long term, the response is positive. For a positive information rush of V_t, the maximum negative response of S_t is achieved in the fourth period. From the seventh period, the response of S_t to V_t becomes positive. These results show that in the short term, the increase in chemical gross industry output value will reduce the accident incidence, but in the long term, the increase will cause the accident incidence to increase. (2) In the short term, the response of accident incidence S_t to investment in safety facilities I_t is positive. In the long term, the response is negative. For a positive information rush of I_t, the maximum positive response of S_t is achieved in the first period. Beginning from the second period, the response becomes negative and continues to strengthen. From the seventh period, the response starts to weaken. That is, in the short term, the increase investment in safety facilities will increase the accident incidence. However, in the long term, the increase in investment in safety facilities will reduce accident incidence. (3) In the short term, the response of accident incidence S_t to per capita wage W_t impulse is positive. In the long term, the response is negative. For a positive information rush of W_t, the maximum positive response of S_t is achieved in the fourth period. Then, the response is negative, and the maximum negative response of S_t is achieved in the ninth period. Namely, in the short term, the increase in per capita wage will cause accident incidence to increase. In the long term, the increase in per capita wage will reduce accident incidence.

From the perspective of the chemical workers' safety awareness, workers tend to generate offsetting behavior, because of workers' behavior in regards to moral hazard. The workers' offsetting behavior is derived from the relatively low knowledge level of chemical workers. The main reason for chemical workers lack knowledge culture is that according to China's family planning policy, each urban family only has one child, and each rural family can have two children. The consequences of the one-child policy are as follows: first, the urban only- child is spoiled and doesn't want to work at the chemical factory in which the working environment is poor; second, the vocational-technical schools in cities designed for chemical plant workers have been forced to close due to lack of students; third, in China, migrant workers[6] have become the current primary labor in high-risk industries such as the chemical industry, especially in private chemical enterprises. Because of migrant workers' education level is relatively low, when judging safety risk, they are more likely to generate offsetting behavior. In the short term, when the enterprises increase investment in safety facilities to provide a safer working environment for workers, the workers will be dependent on enterprise safety precautions excessively, thinking that their working environment is safe, thus reducing their safety awareness[7]. So, workers tend to increase labor efforts to earn higher wages by reducing safety efforts (assume that the worker will allocate his effort between wage effort and safety effort in this paper), leading to an offsetting effect. If the offsetting effect is strong enough, the accident incidence will increase conversely. Therefore, in the short term, with the increase in investment in safety facilities, per capita wage and accident incidence both increase. In the long term, with the improvement of the workers' risk prevention awareness level through strengthening education and training, when increasing investment in safety facilities, workers will be aware of the increase in potential risk, rather than believing that their working environment gets better. Thus, workers will transfer their effort to safety from wages, obtaining a higher safety level. That is, with the workers' offsetting behavior disappearing gradually, an increase in investment in safety facilities will eventually have the effect off reducing accident incidence.

CASE STUDY

Dalian is located in the northeast coastal chemical industry area of China, and the chemical industry is developing rapidly. After the oil pipeline explosion on July 16, 2010 which shocked the world, the nation and the local government attached great importance to chemical safety. In 2010 and 2011, the government passed a total of 6 laws and regulations about hazardous chemicals (30, 32, 36, 40, 41, 42 in order), two times more than the sum of the past five years. Meanwhile, the Dalian government invested 20 million Yuan (investment in safety facilities) to improve the safety level of chemical companies, including the upgrade and reform of enterprise hazardous process automation control, with all the chemical enterprises within the district achieving the process automation control. This 20 million Yuan investment not only achieved the expected effect, but also increased the accident incidence.

On November 25, 2010, a toxic gases spill accident (including carbon oxide, hydrogen sulfide and so on) occurred in Carbon Chemical Co., Ltd. in Dalian, resulting in more than 20 workers, who were doing gymnastics close to the scene, suffering from toxic gas poisoning. The enterprise has sound safety rules and regulations, perfect safety operational procedures and security facilities. The cause of the accident was ash deposits in the gasifier, leading to the stopping of the compressor, and then the system stopped automatically. When the gases in the furnace burned insufficiently, toxic gases emitted directly through the torch. The direct cause of the accident was the workers' over-reliance on the automatic safety chain system. When deviant behavior occurred at the end of the safety chain, there was no human intervention, eventually leading to the gas leakage accident[8].

On August 29, 2011, Dalian Petrochemical Company's refined oil storage tank exploded and caught fire. The incident occurred when the tubing outlet velocity of #875 tank reached 4.34 m/s, due to the floating plate without automatic floating, during the oil delivery operations of refined oil storage tanks. Due to exceeded safety limits, a large amount of static electricity was produced and discharged, igniting the mixture of oil mist, combustible gas and air, which exploded. This was a typical accident caused by violation of rules. When the tubing outlet velocity exceeded the safety limits, the workers didn't control the tubing outlet velocity within safe limits in accordance with the rules, but relied on automatic equipment operation fully, resulting in the storage tank exploding and catching fire.

During the 2010 and 2011, while safety regulation was enhanced, the number[9] of the chemical accidents was 2.7 times more than the total of the previous two years in Dalian. And 62.5% of 2010-2011 accidents belong to the "three violations" category. This indicates that the offsetting behavior is more serious in Dalian. Regulation enhancement did not reduce the accident incidence, but increased the accident incidence, due to workers relying too heavily on security facilities. The main reason is that, several years ago, the original chemical vocational- technical school in Dalian had been forced to close due to insufficient number of students. Migrant workers' safety training is insufficient and safety awareness is weak, which is the reason for serious chemical workers' offsetting behavior in Dalian. The case of Dalian further illustrates workers' offsetting behavior is the main cause of chemical accident increases.

CONCLUSION

This paper adopts the chemical accident incidence, the chemical gross industry output value, investment in safety facilities of chemical industry, and per capita wage of chemical industry employment as the indexes to empirically analyze investment performance of chemical safety facilities using time series data over the period 1981-2011 by VECM. The empirical results indicate that for China 's chemical industry, in the short term, increasing fails to improve the safety level significantly because of the offsetting behavior of the employees. Over the long term, the offsetting behavior tends to diminish, and the chemical accident incidence can be decreased by increasing the investment performance index. Poor safety awareness among workers is one of the causes of accident incidences. Therefore, making sure to heighten the chemical workers' safety awareness is one of the measures to improve the investment performance of chemical safety facilities in the short term.

FUNDING

Social Science Fund Project of Liaoning province (L13BJY033); Social Science Fund Project co-sponsored by province and ministry (13JJD790042).

NOTES

[1]The cause of choosing investment in fixed assets as the investment in safety facilities is that the statistical data of investment in safety facilities not listed separately in China, but included in the investment in fixed assets. Another reason is that good working environments and advanced equipment are the safety guarantees for workers.

[2]Peltzman (1975) found that the increase of auto safety equipment did not reduce traffic mortality in the study of automobile safety regulation effect because of the offsetting effect caused by the behavior of drivers. Klick and Stratmann (2003) defined this effect as the Peltzman effect.

[3]The data is not reported here to conserve space but is available from the author upon request.

[4]The main purpose of this paper is to analyze the impact on accident incidence made by chemical gross industry output value, investment in safety facilities, and per capita wage. Thus, only the equation in which accident incidence is the explained variable is listed.

[5]The main purpose of this paper is to analyze the impact on accident incidence made by chemical gross industry output value, investment in safety facilities and per capita wage. Therefore, only test the equation in which accident incidence is the explained variable.

[6]Refers to the agricultural registered permanent residents working at the local township enterprise or urban enterprise.

[7]Viscusi (1979) through study found that when companies improve workers' working conditions and the quality of labor safety, workers' safety efforts will decline and the action level preventing risk will drop.

[8]See the two cases, Dalian Administration Bureau of Safety Working in China (2010-2011).

[9]The number of chemical accidents is taken from the China Chemical Safety Association.

REFERENCES

1.   Sun, K. and Yang, H.M. (2012) Statistical Analysis of Dangerous Chemical Accidents in China. Fire Technology, 48, 331-341. http://dx.doi.org/10.1007/s10694-011-0224-y

2.   Smith, R.S. (1979) The Impact of OSHA Inspection on Manufacturing Injury Rates. Journal of Human Resource, 14, 145-170.

3.   Gray, W.B. and Scholz, J.T. (1993) Does Regulatory Enforcement Work? A Panel Analysis of OSHA Enforcement. Law & Society Review, 27, 177-213.http://dx.doi.org/10.2307/3053754

4.   Lewis-Beck, M.S. and Alford, J.R. (1980) Can Government Regulate Safety? The Coal Mine Example. The American Political Science Review, 74, 745-756.http://dx.doi.org/10.2307/1958155

5.   Carmichael, H.L. (1986) Reputations for Safety: Mark Performance and Policy Remedies. Journal of Labor Economics, 4, 458-472. http://dx.doi.org/10.1086/298106

6.   Weil, D. (1996) If OSHA Is So Bad, Why Is Compliance So Good? The Rand Journal of Economics, 27, 618-640. http://dx.doi.org/10.2307/2555847

7.   Viscusi, K. (1992) Fatal Tradeoffs: Public and Private Responsibilities for Risk. Oxford University Press, New York, 50-171.

8.   Viscusi, K. (1979) The Impact of Occupation Safety and Health Regulation, 1973-1983. Bell Journal of Economics, 10, 117-140. http://dx.doi.org/10.2307/3003322

9.   Keiser, K.R. (1980) The New Regulation of Health and Safety. Political Science, 95, 479-491. http://dx.doi.org/10.2307/2150061

10. Greenberg, E.S. (1985) Capitalism and the American Political Ideal. M. E. Sharpe, Armonk, 76-80.

11. Reniers, G.L.L., et al. (2008) A Multiple Shutdown Method for Managing Evacuation in Case of Major Fire Accidents in Chemical Clusters. Journal of Hazardous Materials, 152, 750-756. http://dx.doi.org/10.1016/j.jhazmat.2007.07.040

12. He, G.Z., et al. (2011) Managing Major Chemical Accidents in China: Towards Effective Risk Information. Journal of Hazardous Materials, 187, 171-181.http://dx.doi.org/10.1016/j.jhazmat.2011.01.017

13. Zhang, X.M. and Chen, G.H. (2011) Modelin and Algorithm of Domino Effect in Chemical Industrial Parks Using Discrete Isolated Island Method. Safety Science, 49, 463-467.http://dx.doi.org/10.1016/j.ssci.2010.11.002

14. Klick, J. and Stratmann, T. (2003) Offsetting Behavior in the Workplace. Working Paper, George Mason University, Feldstein.

15. Feldstein, M. and Stock, J.H. (1994) The Use of a Monetary Aggregate to Target Nominal GDP. Monetary Policy, University of Chicago Press, Chicago, 7-69.

16. Toda, H.Y. and Phillips, P.C.B. (1993) Vector Autoregressions and Causality. Econometrica, 61, 1367-1393. http://dx.doi.org/10.2307/2951647

Chapter 2

PREPARATION, PHARMACOKINETICS, BIODISTRIBUTION, ANTITUMOR EFFICACY AND SAFETY OF LX2-32C-CONTAINING LIPOSOME

Hongbo Wang[1]., Jianqiao Zhang[1]., Guangyao Lv[1], Jinbo Ma[3], Peng-kai Ma[1], Guangying Du[1], Zongliang Wang[1], Jingwei Tian[1,4], Weishuo Fang[2], Fenghua Fu[1]

[1]. Key Laboratory of Molecular Pharmacology and Drug Evaluation (Ministry of Education of China), School of Pharmacy, Yantai University, Yantai, China

[2]. State Key Laboratory of Bioactive Substances and Functions of Natural Medicines, Institute of Materia Medica, Chinese Academy of Medical Sciences and Peking Union Medical College, Beijing, China

[3]. Department of clinical medicine, Binzhou Medical College, Yantai, China

[4]. State Key Laboratory of Long-acting and Targeting Drug Delivery Technologies (Luye Pharma Group Ltd.), Yantai, China

ABSTRACT

Lx2-32c is a novel taxane that has been demonstrated to have robust antitumor activity against different types of tumors including several paclitaxel-resistant neoplasms. Since the delivery vehicles for taxane, which include cremophor EL, are all associated with severe toxic effects, liposome-based Lx2-32c has been developed. In the present study, the pharmacokinetics, biodistribution, antitumor efficacy and safety characteristics of liposome-based Lx2-32c were explored and compared with those of cremophor-based Lx2-32c. The results showed that liposome-based Lx2-32c displayed similar antitumor effects to cremophor-based Lx2-32c, but with significantly lower bone marrow toxicity and cardiotoxicity, especially with regard to the low ratio of hypersensitivity reaction. In comparing these two delivery modalities, targeting was superior using the Lx2-32c liposome formulation; it achieved significantly higher uptake in tumor than in bone marrow and heart. Our data thus suggested that the Lx2-32c liposome was a novel alternative formulation with comparable antitumor efficacy and a superior safety profiles to cremophor-based Lx2-32c, which might be related to the improved pharmacokinetic and biodistribution characteristics. In conclusion, the Lx2-32c liposome could be a promising alternative formulation for further development.

INTRODUCTION

Paclitaxel has been widely used as a chemotherapeutic agent in the treatment of a broad range of cancers [1]–[3]. However, its clinical usefulness has been limited by drug-resistance and delivery vehicle-related toxic effects, especially regarding the hypersensitivity induced by cremophor EL castor oil [4]–[6]. Therefore, novel taxanes involving cremophor EL free formulations that retain their sensitive regarding paclitaxel-resistant cancers would be of great clinical benefit.

Previously, we reported for the first time on Lx2-32c, a novel taxane semisynthesized from cephalomannine [7], [8]. This compound displayed robust anticancer activity against several cancer cell lines both *in vitro* and *in vivo*, especially against several paclitaxel-resistant lines such as A549/taxol and A2780/paclitaxel [8], [9]. Based on this finding, Lx2-32c has been viewed as a potential candidate to overcome paclitaxel-resistance in the clinic. However, because of its aqueous insolubility, Lx2-32c must be dissolved in the same vehicle (cremophor EL and anhydrous ethanol [1:1 V/V]) as the paclitaxel, which will introduce similar severe side effects to those of taxol [4], [10]. Therefore, an improved formulation, which could eliminate the adverse effects associated with the solvent while retaining similar anticancer activity, would greatly benefit cancer patients and accelerate the development process.

To overcome the solvent challenge regarding taxane, some cremophor-free or reduced cremophor EL paclitaxel formulations, such as liposomes, Genexol- PM, AI850, Genetaxyl or an albumin-bound nanoparticle formulation of paclitaxel, have been approved or developed in clinical trials [11]–[17]. Amongst them, liposomes have been used to encapsulate a variety of pharmacological agents, such as doxorubicin and paclitaxel (Lipusu, paclitaxel liposome for injection) [18], [19]. It has been well documented that liposome encapsulation of doxorubicin could reduce local irritation and vesicant action without a significant decrease in antitumor effect[20]. In addition, paclitaxel when encapsulated in a liposome (Lipusu) instead of a conventional excipient has been shown to have a markedly reduced toxicity while retaining equal efficacy in mice and rat cancer models [18]. Consequently, a liposome-based formulation of Lx2-32c might be effective in cancer treatment without the toxic side effects induced by cremophor-based Lx2-32c. In the present study, the Lx2-32c liposome was prepared and its antitumor effect, toxicity, biodistribution and pharmacokinetics were evaluated and compared with those of cremophor-based Lx2-32c.

METHODS

Chemicals and Animals

Lx2-32c was obtained from the Chinese Academy of Medical Sciences, and the purity of the compound used in the present study was higher than 98% as checked by HPLC. The 100% ethanol and cremophor EL castor oil were kindly supplied by the Beijing Union Pharmaceutical Factory. Lecithin was provided by Avanti Polar Lipids, Inc. Cholesterol was purchased from the Hubei Kangbaotai Fine-Chemicals Co. Ltd.

Male C57BL/6J mice (18–22 g) were provided by the Beijing HFK Bioscience Co., Ltd, and male SD rats (200–240 g) were obtained from the Shandong Luye Pharmaceutical Co., Ltd. The animals were housed in a light and temperature-controlled room (21–22°C; humidity 60–65%) and maintained on a standard diet and water. All of the experimental protocols were approved by Committee on the Ethics of Animal Experiments of Yantai University. All surgery was performed under sodium pentobarbital anesthesia, and all efforts were made to minimize suffering.

Preparation of Lx2-32c Liposomes

Lx2-32c liposomes were prepared using a film dispersion method followed by a lyophilization technique. Briefly, Lx2-32c (60 mg), lecithin (720 mg) and cholesterol (108 mg) were dissolved in chloroform. After 5 min of stirring, the organic solvent was evaporated on a rotary evaporator under reduced pressure at 40°C to obtain a membrane. The resulting membrane was dissolved by the addition of PBS (pH=7.4) to obtain the Lx2-32c liposome solution. The solution obtained using this process was disrupted in a 200-W ultrasonic homogenizer for 20 min. Then, the remaining solution was lyophilized to dried Lx2-32c liposomes using a freeze dryer system (Labconco, USA) and their microstructure was observed using a scanning electron microscope. Particle size was evaluated by means of a particle size analyzer (Mastersizer 2000, Malvern Instruments). The encapsulation efficacy of LX2-32c was measured using high-performance liquid chromatography (HPLC; Shimadzu LC-20A, JPN; C18 column, 250×4.6 mm; 5 μm).

Evaluation of Antitumor Effects and Toxicity in B16 Tumor-Bearing C57BL/6J Mice

C57BL/6 mice were used to establish xenograft tumors of murine melanoma (B16) as previously reported [21]. In this experiment, the tumors were isolated from donor mice and implanted in the dorsum of recipient mice by means of subcutaneous injection. The day after implantation, the animals were randomized into three groups each containing 10 animals: the control group was given a single dose of 0.9% NaCl by intraperitoneal injection; the Lx2-32c cremophor EL group (the compound) was administrated in three doses by intraperitoneal injection in a 10 ml/kg injection volume (30 mg/kg) twice every week; the Lx2-32c liposome group was administrated in three doses by intraperitoneal injection in a 10 ml/kg injection volume (30 mg/kg) twice every week.

Hypersensitivity was assessed according to the grading standards detailed in Table 1 in line with our previous report [22]. Blood samples were collected in tubes containing EDTA for hematological analysis, and in tubes without anticoagulation agent for the evaluation of clinical chemical parameters after the final dose. The weight of the tumor obtained from each mouse was measured to evaluate the antitumor effects. Hematological parameters such as red blood cell (RBC) count, hemoglobin (HGB) level, white blood cell (WBC) count and platelet (PLT) count were compiled at the 407 Naval Hospital of Yantai, China. For the clinical chemistry study, blood samples were centrifuged at 4000 rpm for 10 min at room temperature without anticoagulation. The level of creatine kinase MB (CK-MB) in serum was also assayed at the 407 Naval Hospital of Yantai.

Table 1. Hypersensitivity reactions grading standard

Grade	Clinical signs
0/−	Normal
1/+	Dyspnea, syncope, gatism
2/++	Disturbance, head shaking
3/+++	Shortness of breath, drowsiness
4/++++	Mortality

Pharmacokinetic Experiments in Male SD Rats

The pharmacokinetics of the Lx2-32c liposomes and the Lx2-32c cremophor EL were explored using SD rats after a single intraperitoneal injection at a dose of 30 mg/kg. Briefly, around 500 μl of blood from the posterior orbit was collected into a heparinized vacutainer tube before compound administration and at 0.17, 0.25, 0.5, 1, 2, 4, 8, 12, 24 and 48 h post administration. The blood samples were centrifuged at 8000 rpm for 10 min at 4°C to obtain plasma samples, which were kept frozen at −80°C until analysis. Each plasma sample was treated with 1 ml of acetonitrile and vortexed for 30 s followed by centrifugation at 12000 rpm for 10 min at 4°C. Then, the supernatants were transferred to other test tubes. The Lx2-32c concentration in the plasma samples was quantified using a Diamonsil-C18 (4.6 mm×250 mm; 5 μm) HPLC column at a temperature of 30°C. The mobile phase was 80:20 (methanol:water) at a flow rate of 1 ml/min. The effluent was detected at 225 nm and the area under the peak was used for quantification. Pharmacokinetic parameters were evaluated using Winnonlin Software (Version 6.1, Pharsight Corporation).

Biodistribution of Lx2-32c in B16 Tumor-Bearing C57BL/6J Mice

The mice bearing the xenograft tumor model were separated into Lx2-32c liposome and Lx2-32c cremophor EL groups (40/group) for the biodistribution study. Three animals were chosen from each group at 0.17, 0.25, 0.5, 1, 2, 4, 8, 12, 24 and 48 h post intraperitoneal injection of a 30 mg/kg dose. The hearts and tumors were collected, weighed and frozen at −80°C until assayed. Bone marrow was harvested from femurs using a carefully standardized protocol that involved flushing 500 μl of normal saline through the marrow cavity and collecting the effluent for analysis.

All tissue samples were homogenized in 5 ml acetonitrile. The homogenizer was rinsed with 1 ml acetonitrile to recover the residual drug. The samples were placed on ice and centrifuged for 10 min at 12000 rpm. The supernatants were transferred to other test tubes. The concentration of Lx2-32c in all tissue samples was quantified using the same method and conditions as used for the pharmacokinetic experiments.

Statistical Analyses

The results were presented as mean ± SD. Comparisons between more than 2 groups were performed by analysis of variance (one way ANOVA), then Student t test were performed. $P<0.05$ was used as the level of statistical significance unless indicated otherwise.

RESULTS

Characterization and Properties of the Lx2-32c Liposome

The Lx2-32c liposomes were tested with a mean diameter of 225.9 nm (Fig. 1B), and the average zeta potential was −14.16 mV (Fig. 1C), in which their appearance under scanning electron microscopy was shown (Fig. 1A), The encapsulation efficiency was about 87% and the loading efficiency was 5.3%as determined by HPLC (S1 and S2 Tables in S1 File). No obvious changes were observed on the mean diameter, polydispersity index and encapsulation efficiency for the freeze-dried liposome powder stored in −20°C for at least 10 months (S3 Table in the S1 File).

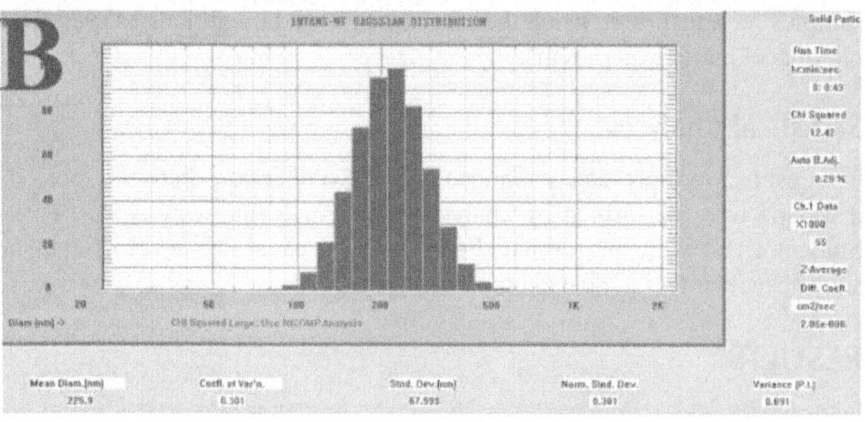

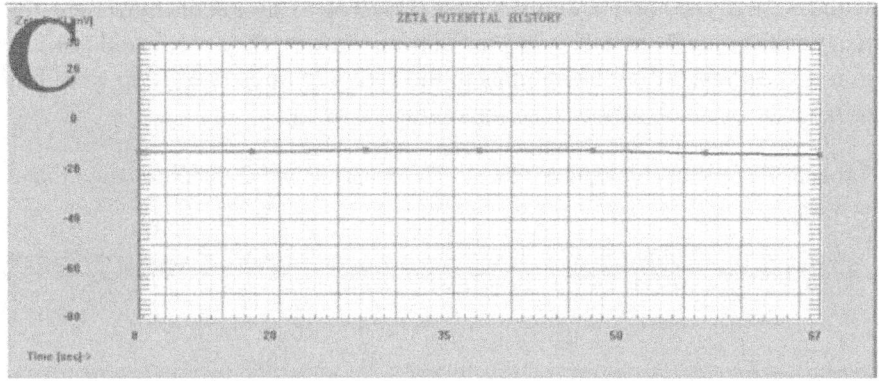

Figure 1. Characterization of the Lx2-32c liposome. A, Scanning electron microscope photograph of Lx2-32c liposome; B, The mean diameter and polydispersity index (PDI) of freshly prepared Lx2-32c liposome; C, The average zeta potential of Lx2-32c liposome.

Antitumor Effect of Lx2-32c Liposome against B16 Xenograft Tumor in C57BL/6J Mice

The antitumor activity was evaluated using xenograft tumor models. The results showed that the Lx2-32c liposome could significantly inhibit the growth of B16 xenograft tumors compared with the control group (Table 2; $P<0.01$); antitumor activity was similar to that in the Lx2-32c cremophor EL group. However, the body weight of animals in the Lx2-32c liposome group was significantly higher relative to that in the Lx2-32c cremophor EL group (Table 2; $P<0.01$).

Table 2. Inhibitory effects of Lx2-32c liposome on the xenograft tumor growth of B16 in C57BL/6J mice

Group	Dosage (mg/kg)	Body weight (g)	Body weight gain (g)	Tumor weight(g)	Inhibition Rate (%)
Control	-	29.44±1.96	4.24±1.53	3.63±0.76	--
Cremophor-based	30	24.00±1.05*	-2.30±1.25*	1.53±1.16*	57.85
liposome	30	27.8±1.99#	2.20±1.87*, #	1.66±0.88*	54.27

Data are expressed as means ± SD (n=10).
*: $p<0.05$, compared with that in control group; #: $p<0.05$, compared with that in Cremophor-based Lx2-32c group.

Hypersensitivity Reactions

The mice were injected intraperitoneally with Lx2-32c in cremophor EL or Lx2-32c liposomes at a dose of 30 mg/kg. Their behavior was subsequently observed and their hypersensitivity reactions were ranked according to the criteria detailed in Table 1. As shown in Table 3, almost all of the animals injected with Lx2-32c cremophor solution were observed to have acute hypersensitivity reactions (grade 3) at 2–5 min post administration; they recovered within 30 min. In contrast, all of the animals in the Lx2-32c liposome group were found to have no or much milder reactions (grade 0 or 1).

Table 3. Hypersensitivity grade of Lx2-32c liposome and cremophor-based Lx2-32c in mice

No.	sterile saline	Cremophor-based Lx2-32c	Lx2-32c liposome
1	–	+++	+
2	–	+++	–
3	–	+++	–
4	–	+++	+
5	–	+++	–
6	–	+++	–

Hematologic Examination

After 2 weeks of administration, the animals in the Lx2-32c cremophor EL group were observed with a markedly reduced WBC count compared with that in the control group (Table 4, $P<0.05$). However, the animals in the Lx2-32c liposome group were found to have a much less pronounced reduction in the WBC count compared with that in the Lx2-32c cremophor group ($P<0.05$). In addition, a significant decrease in the HGB concentration was observed in the animals in the Lx2-32c cremophor group, which was not observed in the Lx2-32c liposome group (Table 4). No significant effect on the RBC and PLT counts was observed in either the Lx2-32c liposome or the Lx2-32c cremophor group (Table 4).

Table 4. The effect of Lx2-32c liposome on WBC, RBC, PLT counts, HGB concentration and CK-MB in C57BL/6J mice

Group	Dosage(mg/kg)	WBC counts (10^9/L)	RBC counts (10^{12}/L)	PLT counts (10^9/L)	HGB (g/L)	CK-MB (U/L)
Control		17.3±1.65	3.2±0.6	364.1±98.6	64.6±10.9	240.0±25.6
Cremophor-based	30	3.1±1.3*	3.2±1.4	460.6±142.2	49.2±5.3*	409.6±86.7*
liposome	30	10.7±5.9*.#	3.2±0.8	3764.8±162.4	65.8±8.6#	269.8±33.8#

Data are expressed as means ± SD (n=10).
*$p<0.05$, compared with that in control group;
#$p<0.05$, compared with that in Cremophor-based Lx2-32c group.

Serum Myocardial Enzyme

CK-MB is a popular biomarker that is used to monitor cardiac injury in the clinic [23]. As shown in Table 4, a marked increase in the activity of CK-MB relative to that in the control group was observed in the animals in the Lx2-32c cremophor group ($P<0.05$); a significant reduction in CK-MB was found in the Lx2-32c liposome group compared with that in the cremophor EL group ($P<0.05$).

Pharmacokinetic Analysis

The pharmacokinetic profiles of Lx2-32c in the liposome and cremophor groups were calculated using the non-compartmental analysis with Winnonlin Software and listed as the mean ± SD in Table 5. As shown in Fig. 2, the plasma drug concentration observed was significantly higher in the Lx2-32c liposome group than that in the Lx2-32c cremophor group ($P<0.01$) at 12 h, 24 h and 48 h post administration, and the accumulation and clearance of Lx2-32c occurred over a considerably shorter period in the Lx2-32c cremophor group compared with the Lx2-32c liposome group.

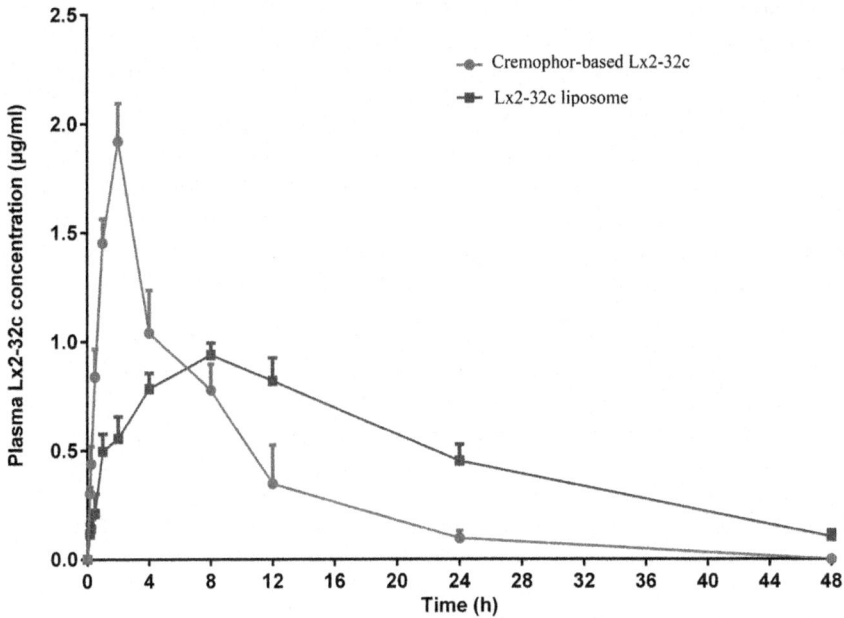

Figure 2. Lx2-32c Plasma concentration–time profile in SD rats following a single i.p. dose of Lx2-32c liposome and cremophor-based Lx2-32c at 30 mg/kg. Data are expressed as mean ± SD (n=4).

Table 5. Mean pharmacokinetic parameters of Lx2-32c liposome and cremophor-based Lx2-32c in SD rats

Parameter	Cremophor-based Lx2-32c	Lx2-32c liposome
$AUC_{(0-t)}$ (mg/L *h)	15.16±1.41	23.37±2.71
$MRT_{(0-t)}$ (h)	7.94±0.71	19.52±0.77
C_{max} (µg/µl)	1.92±0.63	0.94±0.45
CLz/F (L/h/kg)	2.01±0.72	1.19±0.59
Vz/F (L/kg)	16.63±1.07	20.52±0.91

Data are expressed as mean ± SD (n=4).
AUC: Area Under Curve; MRT: Mean Retention Time; CLz/F: Clearance; Vz/F: Apparent Volume of Distribution.

Biodistribution of Lx2-32c

The distribution of Lx2-32c was explored after administration, and the results demonstrated that Lx2-32c reached a peak concentration in the tumor at 12 h after liposome injection and at 2 h after Lx2-32c cremophor injection (Fig. 3). The drug concentration in the tumor was significantly lower in the Lx2-32c liposome group than that in the Lx2-32c cremophor group within the first 4 h post administration ($P<0.01$); it was much higher for the Lx2-32c liposomes than for Lx2-32c cremophor at 8, 12, 24 and 48 h post administration ($P<0.01$). However, the distribution of the drug in the heart tissue (Fig. 3b) and bone marrow (Fig. 3c) differed from that in the tumor; drug uptake in these organs was much high in the Lx2-32c cremophor group relative to the Lx2-32c liposome group. The drug uptake was calculated and the data indicated that the $AUC_{0-48 h}$ was approximately 1.5-fold higher for the tumor, while it was about 2.8- and 1.2-fold lower for heart and bone marrow in the Lx2-32c liposome group compared with that in the Lx2-32c cremophor group (Fig. 4).

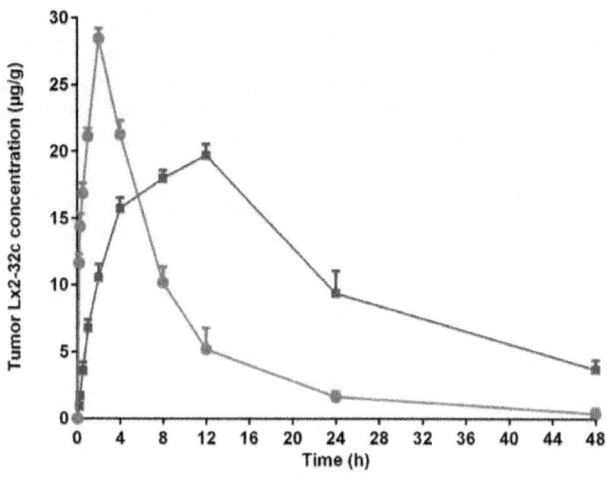

a

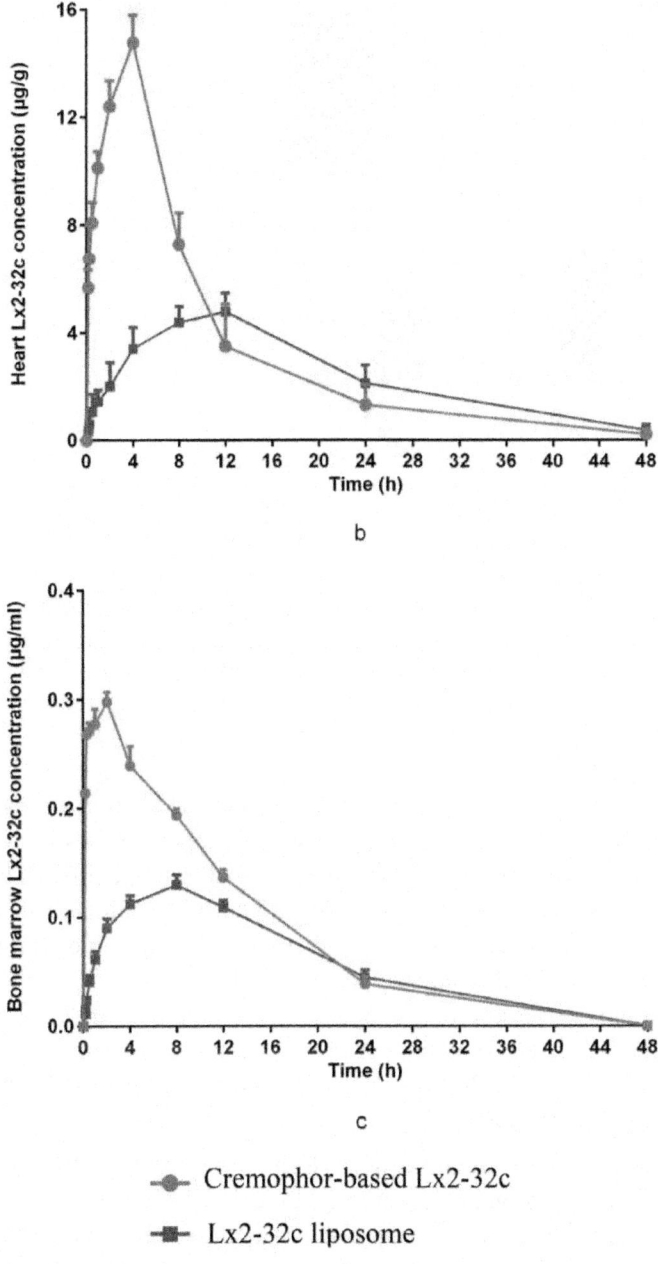

Figure 3. Lx2-32c tissue distribution–time after a single i.p. dose of Lx2-32c liposome and cremophor-based Lx2-32c at 30 mg/kg. (a) tumor; (b) heart; (c) bone marrow. Data are expressed as mean ± SD (n=4).

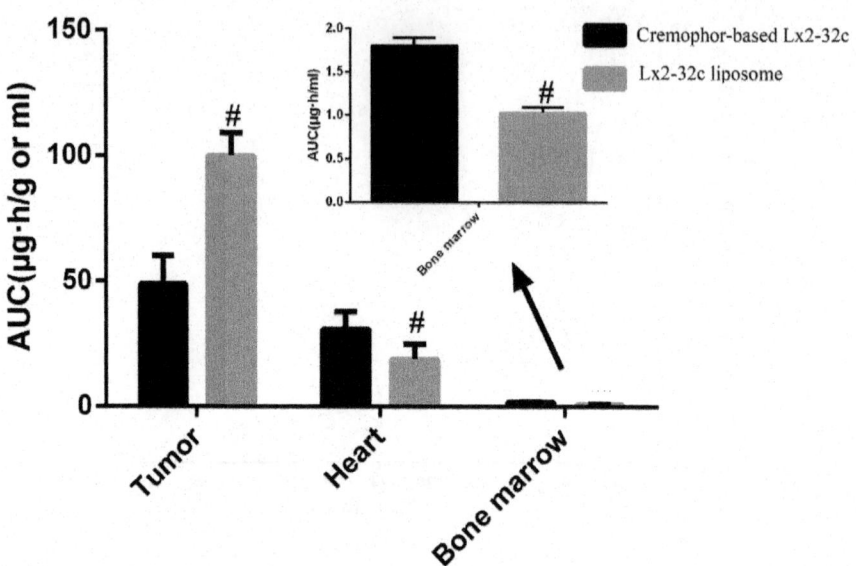

Figure 4. Lx2-32c tissue distribution (AUC) comparative of a single i.p. dose of Lx2-32c liposome and cremophor-based Lx2-32c at 30 mg/kg. Data are expressed as mean ± SD (n=4). #:*P<0.05*, compared with that in the cremophor-based Lx2-32c group.

DISCUSSION

Lx2-32c is a novel taxane derivative semisynthesized from cephalomannine, which was shown to retain sensitivity to paclitaxel in resistant cancer cells both *in vitro* and *in vivo* [8], [9]. To explore an alternative formulation without cremophor, Lx2-32c liposome was developed and evaluated in the current study. Our data demonstrated clearly that Lx2-32c liposome could produce a similar antitumor effect, milder hypersensitivity reactions, lower bone marrow toxicity and lower cardiotoxicity; in addition, it achieved high drug uptake and changed the biodistribution compared with cremophor-based Lx2-32c, which indicated that the Lx2-32c liposome could be an alternative formulation that merits further development.

Taxane is a class of anticancer drugs with robust antitumor activities against different malignancies in the clinic [23]. However, cremophor, the main ingredient in taxane solvents, is always associated with much severer toxic effects, such as hypersensitivity reactions induced by unwanted complement activation [4]–[6]. Indeed, our data showed that almost all the animals injected with cremophor-based Lx2-32c solution were observed to have different degrees of hypersensitivity reactions; these were much milder or not observed in the Lx2-32c liposome group of animals. Based on our previous study, we estimated that the Lx2-32c liposome did not or only lightly induced complement activation compared with cremophor-based Lx2-32c. Importantly, the antitumor activity of Lx2-32c liposomes was similar to that in the cremophor-based Lx2-32c solution when evaluated in terms of reduction in xeograft tumor weight (Table 2). The following toxic effects, however, were much less severe as assayed using change in body weight, in which the animals in Lx2-32c liposomes were observed with obviously improved values of body weight and body weight gain compared with that in cremophor-based Lx2-32c solution group (Table 2).

Bone marrow suppression and cardiotoxicity are the reported common side effects of taxol, which can easily be detected by means of features such as leukopenia and increased expression of biomarkers of myocardial injury, such as CK-MB [24]. In the current study, decreased leukocyte and neutrocyte counts, and increased CK-MB levels as well as bone marrow hypoplasia, were noted in almost all of the animals treated with cremophor-based Lx2-32c; the magnitude of all of these parameters was remarkably reduced in animals injected with Lx2-32c liposomes. All of these findings indicated that this alternative liposome formulation has a much safer toxicity profile than the cremophor formulation. Similar findings have also been reported for liposome-encapsulated doxorubicin and Lipusu, in which the lower toxic effects might be related to lower drug uptake in the target organs [18], [25].

Free drugs, such as paclitaxel and doxorubicin, can be rapidly cleared by absorption through the peritoneal lining and entry into the systemic circulation; this drug clearance pathway can be changed by encapsulation of drugs in liposomes [26], [27]. As a result, liposome-encapsulated drugs may display sustained high concentrations in the immediate vicinity of the target site, and enhanced tumor uptake of the drug. An example is doxorubicin encapsulated in liposomes, which has been shown to be slowly released into the abdominal cavity from disrupted liposomes in a model of peritoneally disseminated cancer [28]. In the present study, higher drug uptake and longer retention time were observed in the tumor tissues of mice injected with Lx2-32c liposomes relative to that injected with cremophor-based Lx2-32c; at the same time the Lx2-32c liposome group was noted as having a much lower peak drug concentration in rat after i.p. injection. Furthermore, all the animals treated with Lx2-32c liposome exhibited a marked delay before the drug entered the systemic circulation; this was partly manifested by a lower distribution and peak concentrations in some normal tissues such as the heart and bone marrow, and much lower subsequent toxic effects. The possible reason for this is that the lymphatic transit of Lx2-32c liposomes from the peritoneum to the systemic circulation was limited to some degree, when cells induced a severe blockage in the lymphatic drainage from the peritoneum [29].

In conclusion, Lx2-32c liposome was evaluated for the first time *in vitro* and *in vivo*. It displayed equal antitumor efficacy to cremophor-based Lx2-32c but considerably lower toxicity (which was related to its lower uptake in normal tissues), and high drug levels in both the circulation system and tumor tissue. Our results thus showed that the Lx2-32c liposome could be an attractive formulation for further evaluation regarding its use in cancer therapy.

SUPPORTING INFORMATION

Table S1. Inter- and intra-day precision of Lx2-32c in plasma

Concentration (μg/ml)	Inter-day (RSD %)	Intra-day (RSD %)
0.1	1.4	3
5	1.9	3.2
200	2.6	3.2

Table S2. Loading efficiency of Lx2-32c liposome formulation

Lx2-32c Liposome Concentration (μg/ml)	Lx2-32c Concentration (μg/ml)	Loading efficiency (%)
5	0.27	5.40
10	0.53	5.30
15	0.80	5.35
35	1.61	5.36
50	2.68	5.36
100	5.22	5.22

Table S3. The mean diameter, polydispersity index (PDI) and encapsulation efficiency (EE) of freshly prepared Lx2-32c liposome and surplus formulation after study

	Mean diameter (nM)	PDI	EE
Fresh	195.5	0.114	87%
Surplus	225.9	0.091	83%

AUTHOR CONTRIBUTIONS

Conceived and designed the experiments: HF HW FF. Performed the experiments: JZ GL PM GD ZW. Analyzed the data: JZ HW JM JT. Contributed reagents/materials/analysis tools: SW WF PM. Wrote the paper: JZ HW.

REFERENCES

1. Wyld L, Reed M (2007) The role of surgery in the management of older women with breast cancer. European journal of cancer 43:2253–2263. doi: 10.1016/j.ejca.2007.07.035

2. Stenner-Liewen F, Zippelius A, Pestalozzi BC, Knuth A (2006) [Molecular targeted therapy]. Der Chirurg; Zeitschrift fur alle Gebiete der operativen Medizen 77:1118–1125. doi: 10.1007/s00104-006-1262-8

3. Fojo T, Menefee M (2007) Mechanisms of multidrug resistance: the potential role of microtubule-stabilizing agents. Annals of oncology: official journal of the European Society for Medical Oncology/ESMO 18 Suppl 5v3–8. doi: 10.1093/annonc/mdm172

4. Szebeni J, Muggia FM, Alving CR (1998) Complement activation by Cremophor EL as a possible contributor to hypersensitivity to paclitaxel: an in vitro study. Journal of the National Cancer Institute 90:300–306. doi: 10.1093/jnci/90.4.300

5. Rowinsky EK, Eisenhauer EA, Chaudhry V, Arbuck SG, Donehower RC (1993) Clinical toxicities encountered with paclitaxel (Taxol). Seminars in oncology 20:1–15.

6. Weiszhar Z, Czucz J, Revesz C, Rosivall L, Szebeni J, et al. (2012) Complement activation by polyethoxylated pharmaceutical surfactants: Cremophor-EL, Tween-80 and Tween-20. European journal of pharmaceutical sciences: official journal of the European Federation for Pharmaceutical Sciences 45:492–498. doi: 10.1016/j.ejps.2011.09.016

7. Wang L, Alcaraz AA, Matesanz R, Yang CG, Barasoain I, et al. (2007) Synthesis, biological evaluations, and tubulin binding poses of C-2alpha sulfur linked taxol analogues. Bioorganic & medicinal chemistry letters 17:3191–3194. doi: 10.1016/j.bmcl.2007.03.026

8. Wang H, Li H, Zuo M, Zhang Y, Liu H, et al. (2008) Lx2-32c, a novel taxane and its antitumor activities in vitro and in vivo. Cancer letters 268:89–97. doi: 10.1016/j.canlet.2008.03.051

9. Zhou Q, Li Y, Jin J, Lang L, Zhu Z, et al. (2012) Lx2-32c, a Novel Taxane Derivative, Exerts Anti-resistance Activity by Initiating Intrinsic Apoptosis Pathway in Vitro and Inhibits the Growth of Resistant Tumor

in Vivo. Biological and Pharmaceutical Bulletin 35:2170–2179. doi: 10.1248/bpb.b12-00513

10. Adams JD, Flora K, Goldspiel B, Wilson J, Arbuck S, et al. (1992) Taxol: a history of pharmaceutical development and current pharmaceutical concerns. Journal of the National Cancer Institute Monographs: 141–147.

11. Crosasso P, Ceruti M, Brusa P, Arpicco S, Dosio F, et al. (2000) Preparation, characterization and properties of sterically stabilized paclitaxel-containing liposomes. Journal of Controlled Release 63:19–30. doi: 10.1016/s0168-3659(99)00166-2

12. Zhang JA, Anyarambhatla G, Ma L, Ugwu S, Xuan T, et al. (2005) Development and characterization of a novel Cremophor EL free liposome-based paclitaxel (LEP-ETU) formulation. European journal of pharmaceutics and biopharmaceutics 59:177–187. doi: 10.1016/j.ejpb.2004.06.009

13. Hamaguchi T, Kato K, Yasui H, Morizane C, Ikeda M, et al. (2007) A phase I and pharmacokinetic study of NK105, a paclitaxel-incorporating micellar nanoparticle formulation. British journal of cancer 97:170–176.

14. Chu Z, Chen J-S, Liau C-T, Wang H-M, Lin Y-C, et al. (2008) Oral bioavailability of a novel paclitaxel formulation (Genetaxyl) administered with cyclosporin A in cancer patients. Anti-cancer drugs 19:275. doi: 10.1097/cad.0b013e3282f3fd2e

15. Straub JA, Chickering DE, Lovely JC, Zhang H, Shah B, et al. (2005) Intravenous hydrophobic drug delivery: a porous particle formulation of paclitaxel (AI-850). Pharmaceutical research 22:347–355. doi: 10.1007/s11095-004-1871-1

16. Mita AC, Olszanski AJ, Walovitch RC, Perez RP, MacKay K, et al. (2007) Phase I and pharmacokinetic study of AI-850, a novel microparticle hydrophobic drug delivery system for paclitaxel. Clinical cancer research 13:3293–3301. doi: 10.1158/1078-0432.ccr-06-2496

17. Kim T-Y, Kim D-W, Chung J-Y, Shin SG, Kim S-C, et al. (2004) Phase I and pharmacokinetic study of Genexol-PM, a cremophor-free, polymeric micelle-formulated paclitaxel, in patients with advanced malignancies. Clinical cancer research 10:3708–3716. doi: 10.1158/1078-0432.ccr-03-0655

18. Ye L, He J, Hu Z, Dong Q, Wang H, et al. (2013) Antitumor effect and toxicity of Lipusu in rat ovarian cancer xenografts. Food and chemical toxicology: an international journal published for the British Industrial Biological Research Association 52:200–206. doi: 10.1016/j.fct.2012.11.004

19. Forssen EA, Tokes ZA (1983) Attenuation of dermal toxicity of doxorubicin by liposome encapsulation. Cancer treatment reports 67:481–484.

20. Madhavan S, Northfelt DW (1995) Lack of vesicant injury following extravasation of liposomal doxorubicin. Journal of the National Cancer Institute 87:1556–1557. doi: 10.1093/jnci/87.20.1556

21. Wang H, Kong L, Zhang J, Yu G, Lv G, et al. (2014) The pseudoginsenoside F11 ameliorates cisplatin-induced nephrotoxicity without compromising its anti-tumor activity in vivo. Scientific reports 4.

22. Wang H, Cheng G, Du Y, Ye L, Chen W, et al. (2013) Hypersensitivity reaction studies of a polyethoxylated castor oil-free, liposome-based alternative paclitaxel formulation. Molecular medicine reports 7:947–952. doi: 10.3892/mmr.2013.1264

23. Rowinsky M, Eric K (1997) The development and clinical utility of the taxane class of antimicrotubule chemotherapy agents. Annual review of medicine 48:353–374. doi: 10.1146/annurev.med.48.1.353

24. Jaffe AS, Babuin L, Apple FS (2006) Biomarkers in acute cardiac diseasethe present and the future. Journal of the American College of Cardiology 48:1–11. doi: 10.1016/j.jacc.2006.02.056

25. Working PK, Newman MS, Sullivan T, Yarrington J (1999) Reduction of the cardiotoxicity of doxorubicin in rabbits and dogs by encapsulation in long-circulating, pegylated liposomes. The Journal of pharmacology and experimental therapeutics 289:1128–1133.

26. Yang A-z, Li J, Xu H-j, Chen H (2006) A study on antitumor effect of liposome encapsulated paclitaxel in vivo and in vitro. Bull Chin Cancer 15:862–864.

27. Rosa P, Clementi F (1983) Absorption and tissue distribution of doxorubicin entrapped in liposomes following intravenous or intraperitoneal administration. Pharmacology 26:221–229. doi: 10.1159/000137805

28. Sadzuka Y, Hirama R, Sonobe T (2002) Effects of intraperitoneal administration of liposomes and methods of preparing liposomes for local therapy. Toxicology letters 126:83–90. doi: 10.1016/s0378-4274(01)00447-7

29. Phillips WT, Medina LA, Klipper R, Goins B (2002) A novel approach for the increased delivery of pharmaceutical agents to peritoneum and associated lymph nodes. The Journal of pharmacology and experimental therapeutics 303:11–16. doi: 10.1124/jpet.102.037119

Chapter 3

IMPROVING THE SAFETY OF STAPHYLOCOCCUS AUREUS POLYVALENT PHAGES BY THEIR PRODUCTION ON A STAPHYLOCOCCUS XYLOSUS STRAIN

Lynn El Haddad[1], Nour Ben Abdallah[2] , Pier-Luc Plante[3] , Jeannot Dumaresq[4] , Ramaz Katsarava[5] , Steve Labrie[6] , Jacques Corbeil[3] , Daniel St-Gelais[2,6], Sylvain Moineau[1]

[1] De´partement de biochimie et de microbiologie, Faculte´ des sciences et de ge´nie, Groupe de recherche en e´cologie buccale, Faculte´ de me´decine dentaire, Fe´lix d'He´relle Reference Center for Bacterial Viruses, Universite´ Laval, Que´bec, Canada

[2] Food Research and Development Centre, Agriculture and Agri-Food Canada, SaintHyacinthe, Que´bec, Canada

[3] De´partement de Me´decine Mole´culaire, Faculte´ de Me´decine, Universite´ Laval, Que´bec, Canada

[4]De´partement de Microbiologie et d'Infectiologie, Centre Hospitalier Affilie´ Universitaire Hoˆtel-Dieu de Le´vis, Le´vis, Que´bec, Canada

[5] Institute of Chemistry & Molecular Engineering, Agricultural University of Georgia, University Campus at Digomi, Tbilsi, Georgia

[6] De´partement des sciences des aliments et de nutrition, Faculte´ des sciences de l'agriculture et de l'alimentation, Dairy Science and Technology Research Centre/ Institute of nutrition and functional foods, Universite´ Laval, Que´bec, Canada

ABSTRACT

Team1 (vB_SauM_Team1) is a polyvalent staphylococcal phage belonging to the *Myoviridae*family. Phage Team1 was propagated on a *Staphylococcus aureus* strain and a non-pathogenic *Staphylococcus xylosus* strain used in industrial meat fermentation. The two Team1 preparations were compared with respect to their microbiological and genomic properties. The burst sizes, latent periods, and host ranges of the two derivatives were identical as were their genome sequences. Phage Team1 has 140,903 bp of double stranded DNA encoding for 217 open reading frames and 4 tRNAs. Comparative genomic analysis revealed similarities to staphylococcal phages ISP (97%) and G1 (97%). The host range of Team1 was compared to the well-known polyvalent staphylococcal phages phi812 and K using a panel of 57 *S. aureus*strains collected from various sources. These bacterial strains were found to represent 18 sequence types (MLST) and 14 clonal complexes (eBURST). Altogether, the three phages propagated on *S. xylosus* lysed 52 out of 57 distinct strains of *S. aureus*. The identification of phage-insensitive strains underlines the importance of designing phage cocktails with broadly varying and overlapping host ranges. Taken altogether, our study suggests that some staphylococcal phages can be propagated on food-grade bacteria for biocontrol and safety purposes.

INTRODUCTION

Staphylococcus aureus is one of the main causes of hospital associated infections [1] and foodborne contaminations [2]. Despite being found on the skin and in mucous membranes of healthy carriers, *S. aureus* is responsible for a wide range of diseases, including mild to severe skin infections, sepsis, endocarditis, and other life-threatening infections. Methicillin-resistant *S. aureus* (MRSA) are often found in a hospital or a community outbreak and their emergence is becoming a global concern [3]. In addition, some *S. aureus* strains cause food poisoning which results from the ingestion of staphylococcal enterotoxins secreted during growth in foods [2]. To combat the detrimental effects of staphylococcal growth, phages are being investigated as an alternative strategy.

However, certain staphylococcal phages encode virulence genes such as Panton-Valentine leukocidin, exfoliative toxin, and enterotoxin genes [4]–[8]. In fact, many temperate phages are a source of lysogenic conversion as they can integrate into the bacterial genome upon infection through the expression of lysogeny-related genes. When phages exist in a prophage state, an induction phenomenon can reactivate and excise their genome from the bacterial chromosome to initiate a lytic cycle. Temperate phages can ultimately play a role in the horizontal transfer of virulence factors from a donor host to a recipient cell [9]. Therefore, it is more suitable to use only virulent phages in a biocontrol application in order to ensure the absence of genes coding for unwanted traits.

To avoid the transfer of virulence genes, the use of a non-pathogenic bacterial strain or a "surrogate" to propagate these phages is preferable, as long as it does not induce changes in the characteristics of the phage. For example, the anti-*Listeria monocytogenes* phages are propagated on strains of non-pathogenic *Listeria innocua* [10]–[12]. Others showed that the propagation of polyvalent *Salmonella* phage phi PVP-SE1 on a non-pathogenic *Escherichia coli* strain did not change the microbiological properties and the DNA restriction profile of that phage. The availability of such non-pathogenic phage-production hosts will facilitate the purification process leading to a safer phage product [13].

Recently, the use of virulent staphylococcal phages as biocontrol agents has been further investigated prompted by the increasing emergence of strains resistant to antibiotics. Some polyvalent staphylococcal phages have been isolated and characterized in the last decade[14]–[19]. Of particular interest are the staphylococcal phages belonging to the *Myoviridae* family (dsDNA genome, icosahedral capsid, and contractile tail) known for their broad host range [20]–[23] and their ability to infect coagulase positive and negative staphylococci (CoPS and CoNS) of animal and human origins [22], [24].

In this study, a new polyvalent phage called Team1 (vB_SauM_Team1) belonging to the*Myoviridae* family was characterized following propagation on a CoPS strain of *S. aureus* and on a non-pathogenic CoNS strain of *Staphylococcus xylosus*. *S. xylosus* is a coagulase-negative staphylococcal species that has long been used as starter culture in Greek, Spanish, Italian and other traditional fermented sausage processes [25]. *S. xylosus* contributes to the aroma and color stability of the final product by minimizing rancidity and, in some cases, providing a bioprotective effect against microbial contamination [26], [27]. We also compared the microbiological and genomic properties of phage Team1 with those of two well-known polyvalent staphylococcal phages K and 812. In parallel, we genotyped a wide range of *S. aureus* strains isolated from different sources and verified the polyvalent nature of the three phages.

MATERIALS AND METHODS

Bacterial Strains and Media

The CoPS strain *S. aureus* SA812 and CoNS strain *S. xylosus* SMQ-121 were used to propagate the phages (Table 1). *S. xylosus* SMQ-121 is a commercially available meat starter. The strains were obtained from the Félix d'Hérelle Reference Center for Bacterial Viruses (www.phage.ulaval/ca/). We established the host ranges of the phages using fifty-six additional *S. aureus* strains isolated from different sources. Tryptic Soy Broth (TSB) was used for culture of all staphylococcal cells.

Table 1. Genotyping of the 57 *S. aureus* strains used in this study

Strains	Source of isolation	ST (CC)	References
SA812	Unknown (Czech Republic)	ST30 (CC30)	[23]
HER1101	Unknown	ST707 (CC47)	This study
HER1049	Unknown	ST25 (CC25)	
HER1225	Unknown	ST9 (CC9)	
A170	Infected wounds (Italy)	ST45 (CC45)	[19]
SMQ1281, SMQ1282	Raw cheese samples (MAPAQ, Canada)	ST352 (CC97)	This study
SMQ1283 to SMQ1299	Mastitis infections (CBMRN, Canada)	ST352 (CC97)	
SMQ1300, SMQ1301	Mastitis infections (CBMRN, Canada)	ST2187 (CC97)	
SMQ1302 to SMQ1319	Mastitis infections (CBMRN, Canada)	ST151 (CC151)	
SMQ1320	Mastitis infections (CBMRN, Canada)	ST351 (CC151)	
SMQ1321	Mastitis infections (CBMRN, Canada)	ST126 (CC126)	
SMQ1322	Mastitis infections (CBMRN, Canada)	ST2270 (CC126)	
CMRSA1	Health-care associated (HA)	ST45 (CC45)	[3]
CMRSA2	Health-care associated (HA)	ST5 (CC5)	
CMRSA3	Health-care associated (HA)	ST241 (CC8)	
CMRSA4	Health-care associated (HA)	ST36 (CC36)	
CMRSA5	Health-care associated (HA)	ST8 (CC8)	
CMRSA6	Health-care associated (HA)	ST239 (CC239)	
CMRSA7	Community acquired (CA)	ST1 (CC1)	
CMRSA8	Health-care associated (HA)	ST217 (CC22)	
CMRSA9	Health-care associated (HA)	ST8 (CC8)	
CMRSA10	Community acquired (CA)	ST8 (CC8)	

doi:10.1371/journal.pone.0102600.t001

Bacterial Strain Identification and Genotyping

The staphylococcal species of all strains was first confirmed through 16S analysis and the API Staph kit (BioMérieux™). We amplified the 16S ribosomal RNA gene by extracting the bacterial genomic DNA [28] and performing a Polymerase Chain Reaction using universal primers 27 forward and 1492 reverse, carried out with a PTC-200 machine (MJ Research, Peltier thermal cycler) as follows: 1 hold at 94°C for 3 min then 35 cycles of 94°C for 1 min, 58°C for 1 min, and 73°C for 1 min 45 s. A final elongation step at 73°C for 5 min was performed after the final cycle. The PCR products were sequenced at the genomic platform of the Centre Hospitalier de l'Université Laval using an ABI Prism 3100 apparatus (Applied Biosystems, Foster City, CA). DNA sequences were compared with sequences available in the GenBank database from the National Center of Biotechnology Information (NCBI) using BLAST analysis (http://www.ncbi.nlm.nih.gov/BLAST/) as well as the Ribosomal Database Project [29]. Furthermore, we genotyped all S. aureus strains using the MLST method [28]. This method targets the sequencing of seven housekeeping genes of ~450 bp. The PCR reactions consisted of: 5-min denaturation at 95°C, followed by 30 cycles of annealing at 55°C for 1 min, extension at 72°C for 1 min, and denaturation at 95°C for 1 min, followed by a final extension step at 72°C for 5 min and sequencing of the PCR products. Each gene sequence obtained was given an appropriate allele number. The combination of the 7 alleles defined the allelic profile, which corresponded to its Sequence Type (ST). This was identified using the MLST database website [28]. Moreover, the different STs acquired were clustered into clonal complexes (CC) using the eBURST algorithm. This algorithm is programmed to assemble the STs that have 6 or 7 alleles in common under the same CC [30].

Phage Preparations

Phage Team1 was isolated as a virulent phage in the Republic of Georgia following a hospital staphylococcal infection [31], [32]. Phage Team1 was propagated for four passages in TSB medium at 37°C on its host strain *S. aureus* SA812 (i.e., Team1-SA812) as well as on *S. xylosus* SMQ-121 (i.e., Team1-SMQ121). Two well-known polyvalent staphylococci phages, phi812 [23] and K [22] were also propagated on both of these strains. The phage preparations were named 812-SA812, 812-SMQ121, K-SA812, and K-SMQ121 for phages phi812 and K, respectively. To propagate the phages, bacteria were grown at 37°C to an optical density at 600 nm of 0.1, and then approximately 10^5 phages were added to the medium. Incubation continued at 37°C until complete bacterial lysis, and the resulting lysate was filtered using a 0.45-µm syringe filter.

Electron Microscopy

A 1.5-ml sample of phage lysate (titer of at least 10^9 PFU/ml) was centrifuged at 23,500×g for 1 h at 4°C. The supernatant was removed, leaving approximately 100 µl in the tube. The phage pellet was washed twice with 1.5 ml of ammonium acetate (0.1 M, pH 7.5). The residual volume (100 µl) was used to prepare the observation grid as follows: 10 µl of the staining solution (2% phosphotungstic acid, pH 7.0) was deposited on a Formvar carbon-coated grid (200 mesh; Pelco International). After 30 s, 10 µl of the washed lysate was mixed with the stain by pipetting up and down. After 90 s, the residual liquid was removed from the grid by touching the edge with blotting paper. Phages were observed at 80 kV using a JEOL 1230 transmission electron microscope available at the Institut de Biologie Intégrative et des Systèmes (IBIS) of the Université Laval.

Microbiological Assays

The one-step growth curve assays of Team1-SA812 and Team1-SMQ121 were carried out in triplicate as previously described [33] with a multiplicity of infection (MOI) of 0.05 and at a temperature of 37°C. The burst size was calculated by dividing the average phage titer after the exponential phase by the average titer before the infected cells began to release virions [33]. Phage counts of Team1, phi812, and K propagated on both species, expressed as EOP values, were obtained using a double-layer plaque titration method. In brief, 5 µl of a phage preparation (undiluted lysate and serially diluted (10^{-1} to 10^{-8})) were spotted on TSB soft agar containing 100 to 200 µl of a staphylococcal culture previously grown overnight. Two biological and two technical repetitions were done for each phage and strain. The EOP values were calculated by dividing the titer of the phage on the tested strain by the phage titer on its host strain. In total, 58 strains (57 *S. aureus* and 1 *S. xylosus*) were tested against the three phages and their derivatives.

Phage DNA Preparation and Sequencing

Phage Team1-SA812, Team1-SMQ121, 812-SA812, 812-SMQ121, K-SA812, and K-SMQ121 genomic DNAs were isolated using a Lambda Maxi kit (Qiagen) with modifications reported elsewhere [34]. The EcoRV (Roche Diagnostics) restriction profile of the six phages were compared to confirm differences between phages Team1, phi812, and K, as well as identities with their respective derivatives amplified on the other strain. The DNA fragments were separated in a 0.8% agarose gel, stained with EZVision (Amresco), and photographed under UV illumination. The sequencing was performed using pyrosequencing technology on a 454 FLX instrument available at the Plateforme d'analyses génomiques of the IBIS. Reads were assembled into a single contig with 32-fold coverage for Team1-SMQ121 and 86-fold coverage for Team1-SA812. For phages phi812-SA812 and phi812-SMQ121, reads were assembled into a single contig with 32-fold and 71-fold coverage, respectively. Finally, for phages K-SA812 and K-SMQ121, the coverage was 72-fold and 62-fold coverage, respectively.

Bioinformatic Analysis

The genomes were analyzed using Staden [35] and BioEdit 7.0.9.0 [36]. Open reading frames (ORF) were identified using the ORFinder tool [37] and GeneMarkS [38]. Each ORF begins with a starting codon (AUG, UUG or GUG) and most are preceded by a Shine-Dalgarno sequence specific to staphylococci and optimally placed approximately 10 nucleotides upstream of the start codon [39]. The putative function of each protein was deduced by homology searches using blast2GO [40]. The theoretical isoelectric point (pI) and the molecular mass (MM) of each deduced phage protein was obtained using Compute pI/Mw available on the ExPASy Web page (http://ca.expasy.org/tools/pi_tool.html). Finally, tRNAs were identified using the tRNAscan-SE server [41] and the ARAGORN program [42]. The genome of phage Team1 was compared with the database on NCBI using BLAST. The two phages sharing the highest nucleotide similarity with phage Team1 were selected. The genomes of these three phages were compared at the DNA level and results were presented in circular alignments using Circos software [24]. Codon usage was determined using the codon usage tool accessible through the DNA 2.0 Web server (DNA 2.0, Menlo Park, CA) and the Countcodon program available on the Kazusa DNA Research Institute Web page (http://www.kazusa.or.jp/codon/). The percentages of synonymous codon usage were calculated for each amino acid. The bacterial codon usage of *S. aureus* JH1 was obtained from the Kazusa DNA Research Institute database for comparison purposes.

Nucleotide Sequence Accession Number

The genome sequence of *S. xylosus* Team1-SMQ121 phage was deposited in GenBank under accession number KC012913.

RESULTS AND DISCUSSION

Growth Curves Comparison Of Phages Team1-SA812 and Team1-SMQ121

Team1 has an icosahedral capsid with a diameter of 90.4±4.1 nm and a long contractile sheath 227.2±7.6 nm in length and 21.6±1.5 nm in width, a morphology resembling other staphylococcal *Myoviridae* [21], [23] (Fig. 1). Before propagating Team1 on *S. xylosus*, the species identity was confirmed using 16S rRNA analysis and API-Staph kit detection. Strain SMQ-121 had 99% identity with *S. xylosus* strains through 16S rRNA analysis and 96.7% identity using the API-Staph kit, confirming that this strain belongs to the *S. xylosus* species.

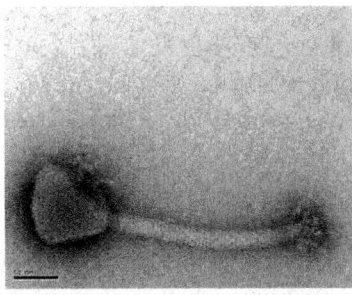

Figure 1. Electron micrograph of phage Team1. The bar scale =50 nm.

The impact of propagating phage Team1 on both *S. aureus* and *S. xylosus* species was determined by comparing their respective growth curves (Fig. 2). The burst size of Team1 was similar on both hosts, with 31±7 PFU per infected *S. aureus* SA812 cell and 37.5±3 pfu per infected *S. xylosus* SMQ-121 cell after approximately 35 minutes. The latency periods (defined as the time interval between the absorption and the beginning of the first burst) were also identical at approximately 20 minutes when grown at 37°C. These values are similar to other*Myoviridae* staphylococcal phages (20, 21, 23). The myophage phi812 that has a longer latent phase and a lower burst size [23] and the newly isolated phages SAH-1 and Stau2 that have similar latent periods but larger burst sizes of 100 pfu per infected cell [20], [21].

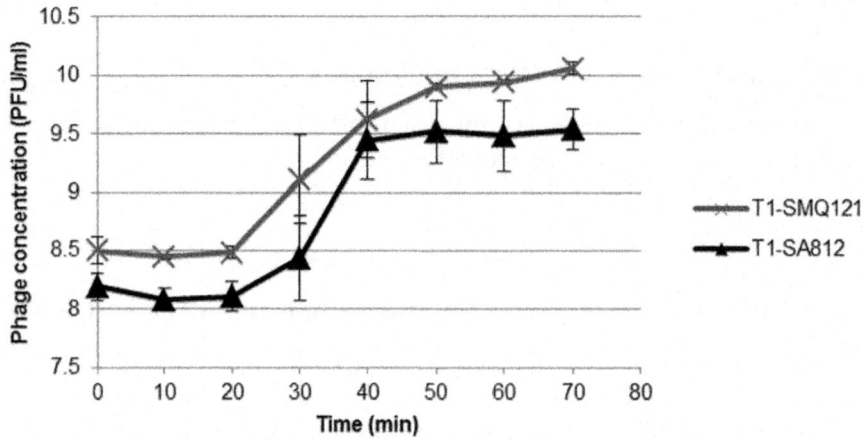

Figure 2. One-step growth curves of both Team1-SA812 (▲) and Team1-SMQ121 (×). Error bars indicate the standard deviation for three trials. The first phage count (time zero) occurred approximately 10 min after adding the phage to the cells [33].

Genome Analyses

Phages Team1-SA812 and Team1-SMQ121 DNA were extracted, sequenced, and compared. The complete genome of Team1 is a 140,903 bp, linear double-stranded DNA. In addition, the Team1 genome has a GC content of 30.3%, which is lower than the estimated GC contents of*S. aureus* and *S. xylosus*, which have GC contents of 32.8% and 34.2%, respectively [25], [43],[44]. However, this GC content is similar to other staphylococcal *Myoviridae* phages [45]. The ends of the phage Team1 genome contain 8,053 bp of long direct terminal repeats (LTRs) separated from the non-redundant part of the virion DNA by 12 bp inverted repeat sequences 5′-TAAGTACCTGGG- 3′ and 5′-CCCAGGTACTTA-3′, which are characteristic features of Twort-like viruses [45]. Homologous recombination between the LTRs permits circularization of infecting phage DNA. Phage Team1 encodes 217 predicted ORFs, 44 of which have a putative function (20%) (Table S1). As with most phages, the Team1 genome can be divided into several modules, including two DNA replication modules and two lysis modules separated by DNA packaging and morphogenesis modules. This modular organization is common among phages belonging to the *Myoviridae* family [8]. The Team1 genome also encodes four tRNAs. The tRNA^{Met} (ATG) is located between ORF33 and ORF34, whereas the three others (tRNA^{Trp}- TGG, tRNA^{Phe} - TTC and tRNA^{Asp} - GAC) are located between ORF76 and ORF77 in proximity to a lysis module. No known virulence factors were found in the phage Team1 genome establishing its safety. Moreover, no lysogenic module or known integrase-related genes were found in its genome, confirming its strictly virulent nature. Finally, comparative analysis of restriction profiles (EcoRV) and genome sequences indicated 100% identity between phages Team1-SA812 and Team1-SMQ121, showing that propagating this phage on two distinct staphylococcal species did not lead to genomic changes.

Comparative Genomics

The genome of phage Team1 was found to be 97% identical to the published genomes of staphylococcal phages ISP [18] as well as G1 [39]. In fact, most staphylococcal myophages, except phage Twort, share identity at the DNA level between each other, ranging from 88.3% to 99.9% [45]. When comparing phage Team1 with G1 and ISP genomes, deletions of 375 bp and 379 bp were observed in the Team1 genome at positions 7,535 and 39,917 respectively (Fig. 3). Deletions were flanked by direct repeat sequences, suggesting that the deletions may have occurred through intramolecular recombination between direct repeats of homologous DNA. The second deletion corresponded to non-coding regions. However, the 375-bp deletion led to the removal of a gene coding for a putative DNA terminal protein, which may explain the shorter LTR sequence in Team1 genome as compared to that of phages G1 and ISP [45]. Insertions of 1,348 bp, 1,375 bp, 146 bp (compared to phage G1), and 260 bp (compared to phage ISP) were also identified at positions 42,626,111,668, and 140,903 in Team1 genome[24]. Two genes were added with these insertions, *orf86* and *orf155*. The former is a hypothetical gene whereas *orf155* encodes a putative intron-encoded endonuclease [46]. The last insertions constituted repetitive non coding sequences.

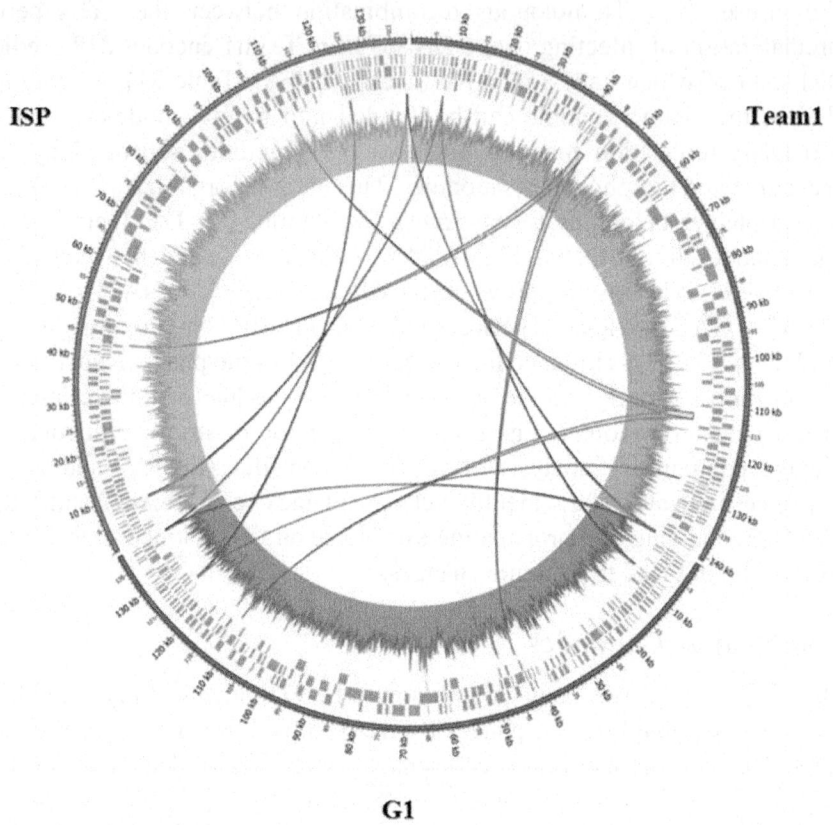

Figure 3. Circular genome comparison of phages Team1, G1, and ISP. Phages Team1, G1, and ISP genomes and GC content are represented, respectively, using the colors purple (upper right), red (center), and green (upper left). For each phage genome, coding regions in minus (blue) and plus (orange) strands are shown in their respective frames. Yellow lines show the additional acquired sequences between these three genomes. The width of the line depends on the length of the acquired sequences [24].

The three phages possess the same four tRNAs including two matching the unique codons for methionine and tryptophan. We also investigated phage codon usage [47] and compared it to *S. aureus* codon usage obtained from the Countcodon program. The codon usage is highly similar between the three phages. While the four tRNAs encoded by the phages are similar to the host tRNAs, significant differences in codon usage were noted between these myophages and the *S. aureus* strain analyzed (Table 2), which likely favors specific phage protein production [48].

Table 2. Codon usage of *S. aureus* **JH1 and phages G1 and ISP for the amino acids encod**ed by the Team1 tRNAs (in bold)

Amino-acid	Codon	Team1	G1	ISP	S. aureus JH1
Cys	TGT	74.0	67.0	66.0	80.6
	TGC	26.0	33.0	34.0	19.4
Asp	**GAC**	**28.0**	**28.0**	**27.0**	**22.0**
Glu	GAG	34.0	32.0	31.0	16.4
Phe	**TTC**	**30.0**	**32.0**	**34.0**	**27.2**
Gly	GGC	7.0	8.0	9.0	15.4
Ile	ATA	46.0	41.0	41.0	22.0
	ATT	39.0	43.0	42.0	60.6
Lys	AAG	38.0	37.0	36.0	19.0
Leu	TTA	40.0	37.0	38.0	59.0
	CTA	19.0	18.0	17.0	9.5
Met	**ATG**	**100.0**	**100.0**	**100.0**	**100.0**
Pro	CCA	28.0	32.0	31.0	50.6
Gln	CAA	62.0	64.0	63.0	87.9
Arg	AGG	24.0	26.0	27.0	4.3
	AGA	59.0	57.0	58.0	33.7
	CGT	7.0	7.0	5.0	37.7
Thr	ACC	18.0	20.0	21.0	4.6
Trp	**TGG**	**100.0**	**100.0**	**100.0**	**100.0**
End	TAA	49.0	51.0	50.0	73.5

doi:10.1371/journal.pone.0102600.t002

Genotyping of *S. aureus* Strains

In the next step, we determined the host range of phages Team1-SA812 and Team1-SMQ121. Before comparing the host ranges of these phages, we collected a panel of 57 *S. aureus* from various sources such as raw cheese, mastitis infections, clinical and hospital associated infections, and community acquired infections. We typed these hosts to assess the diversity of this bacterial set of strains. The typing was done using the MLST method, assigning each strain an ST number, and clustering them into CCs using the eBURST algorithm (Table 1).

The genotypes of the strains differed according to the origin of isolation. *S. aureus* strains isolated from raw cheese or from mastitis infections were found to share some similarities in their CC groups. *S. aureus* is the most common cause of bovine mastitis and often found in raw milk [49]. Mastitis is characterized by the inflammation of mammary glands inducing bacterial infections that result in global economic losses to the dairy industry. The studied *S. aureus* strains isolated from milk could be divided into six groups according to their ST numbers (352, 151, 351, 2270, 2187, and 126). However, when applying the eBURST algorithm and assembling STs sharing at least 6 alleles out of 7, the *S. aureus* strains isolated from milk were divided into three groups or three CCs, i.e., CC97, CC126, and, CC151. The majority of these strains belongs to CC97 (47.5%) and CC151 (47.5%). Other studies showed that the latter CCs are distributed worldwide and constitute the major groups causing mastitis infection outbreaks, mainly in North and South America and Europe [50], [51]. However, CC126 is most predominant in Brazil, along with CC97 [52]. Unlike CC97, CC151 and CC126 are exclusively confirmed in bovine isolates. It is also hypothesized that bovine strains worldwide derived from CC97. The CC97 lineage has existed for more than 50 years and emerged earlier than CC151[50].

The other strains, whether MSSA (methicillin-sensitive *S. aureus*) or MRSA, had different STs and CCs, demonstrating source-dependent genotypes. The MSSA strains, HER1101, HER1049, HER1225, A170, and the host strain SA812, belong to CC707, CC25, CC9, CC45, and CC30 respectively. Conversely, the Canadian MRSA (CMRSA), hospital and community-acquired, have their own separate clustering and belong to different clonal complexes, i.e., CC1, CC5, CC8, CC22, CC36, CC45, CC239. Two clonal complexes share some similarities between the MSSA and MRSA strains studied. Strains CMRSA1 and A170 belong to CC45. Furthermore, CC30 (*S. aureus* SA812) and CC36 (CMRSA4 relevant to EMRSA-16 and USA200) have only two loci of difference in their allelic profile. Additionally, virulent MRSA strains belonging to CC1, CC5, and CC8 are responsible for several outbreaks in North America [3], [53] and are shifting to animal hosts as well [54].

Host Range Comparisons of Phages

After typing the 57 *S. aureus* strains, the host ranges of phages Team1-SA812 and Team1-SMQ121 were examined. We also tested the host ranges of the known polyvalent staphylococcal phages phi812 and K to compare with Team1. These two phages were also amplified on *S. aureus* SA812 and *S. xylosus* SMQ-121 and their genome sequenced to confirm their identity. No mutation was found when amplified on both hosts, as observed for Team1 (data not shown).

The results of the host ranges of the wild-type phages Team1, phi812, and K propagated on both *S. aureus* SA812 and *S. xylosus* SMQ-121 are reported in Table 3. Each of the three phages propagated on *S. aureus* SA812 infected 52 of the 57 *S. aureus* tested. In fact, 56 out of the 57 strains could be infected by at least one myophage. Only strain 1049 (ST25, CC25) was resistant to all three myophages. Of note, this strain is sensitive to *Podoviridae* phages P68 and 44AHJD (data not shown). When the same phages were propagated on *S. xylosus* SMQ-121, phages Team1 and phi812 infected 53 strains, while phage K infected 51 strains. Overall, 55 out of the 57 strains could be infected by at least one myophage propagated on *S. xylosus* SMQ-121. Only the phage sensitivity of strain SMQ1321 (ST126, CC126) was affected by the propagating host, suggesting the presence of host factors that modify the phage behavior in this unique case [48].

Table 3. Host range of phages Team1, phi812, and K propagated on *S. aureus*SA812 and on *S. xylosus* SMQ-121

		S. aureus SA812			*S. xylosus* SMQ-121		
		Team1	phi812	K	Team1	phi812	K
	SA812	1.0	1.0	1.0	1.0	1.6	6.7
	SMQ-121	0.2	1.6	1.3	1.0	1.0	1.0
Other	1101	1.0	0.3	0.3	0.3	0.2	0.3
strains	1049	0	0	0	0	0	0
	1225	0.9	0.1	9.4E-2	0.5	0.3	0.7
	A170	1.4	3.8E-2	0	5.7	1.9E-2	0.7
Raw	SMQ1281	2.5	3.2	1.4	1.0	0.3	1.2
milk	SMQ1282	2.8	4.5	1.9	6.1	0.4	0.5
	SMQ1283	1.3	1.1	1.5	2.8	3.6	6.7
	SMQ1284	2.3	1.8	2.1	1.6	4.0	10.0
	SMQ1285	1.4	1.6	1.7	2.6	2.7	13.0
	SMQ1286	1.4	2.7	1.8	1.6	2.7E-2	3.3E-3
	SMQ1287	1.0	0.7	1.9	1.2	5.7	9.2
	SMQ1288	1.3	1.3	1.6	0.7	3.7	5.0
	SMQ1289	1.5	2.2	1.5	0.9	1.3	8.3
	SMQ1290	1.2	2.0	1.5	1.9	3.6	11.0
	SMQ1291	1.5	1.8	1.4	2.6	6.0	7.5
	SMQ1292	7.7E-7	1.1E-5	6.0E-6	2.3E-4	4.8E-5	0
	SMQ1293	1.9	1.6	1.7	2.8	8.4	11.0
	SMQ1294	1.7	0.9	1.0	1.9	6.0	6.7
	SMQ1295	1.3	1.3	0.9	2.0	0.4	1.1
	SMQ1296	2.4	3.1	4.1	1.3	1.2	1.6
	SMQ1297	2.5	1.3	1.0	0.9	0.9	3.8
	SMQ1298	0.3	1.5	0.7	3.0	0.7	0.2
	SMQ1299	0.5	2.3	1.8	1.5	0.6	1.3
	SMQ1300	0.5	1.7	0.8	1.5	0.3	0.9
Mastitis	SMQ1301	0.2	1.8	0.2	2.0	0.6	0.5
	SMQ1302	4.0	2.3	0.4	0.9	0.8	1.2
	SMQ1303	0.5	2.0	1.4	1.7	0.5	3.3
	SMQ1304	2.0	1.6	2.9	1.9	7.2	9.2
	SMQ1305	1.7	1.6	1.4	2.1	8.4	7.5
	SMQ1306	2.4	2.9	2.0	2.3	3.3	11.0
	SMQ1307	0	0.4	0.6	0	0.5	0.7
	SMQ1308	2.0	1.8	1.6	1.9	2.5	18.0
	SMQ1309	2.4	3.7	5.9	1.4	2.2E+2	41.0
	SMQ1310	1.6	3.1	0.4	1.4	6.1	19.0
	SMQ1311	1.9	2.7	1.6	2.1	1.9	14.0
	SMQ1312	2.4	2.7	1.5	3.0	4.8	10.0
	SMQ1313	1.6	2.5	1.3	3.5	6.0	6.7
	SMQ1314	0.8	1.6	0.9	2.8	12.0	7.5
	SMQ1315	0.8	2.5	1.4	1.4	9.6	5.0
	SMQ1316	1.1	2.7	1.5	1.4	13.0	11.0
	SMQ1317	1.5	1.5	1.4	2.6	12.0	9.2
	SMQ1318	1.5	3.8	1.9	2.4	0.7	0.4
	SMQ1319	0.1	2.5	2.2	2.0	1.0	0.7
	SMQ1320	2.5	8.2	4.4	2.8	0.9	1.9
	SMQ1321	1.9E-5	3.6E-6	2.2E-6	0	0	0
	SMQ1322	1.5E-4	4.0E-4	6.0E-5	1.4E-3	1.4E-4	0

		S. aureus SA812			S. xylosus SMQ-121		
		Team1	phi812	K	Team1	phi812	K
	MRSA1	1.6	0.1	3.8	1.4	7.5E-2	1.3E-5
	MRSA2	0	1.2	0.8	0	0.2	0.4
	MRSA3	1.9	2.7E-2	3.1E-7	1.6	4.5E-3	1.3E-5
	MRSA4	2.2	1.1	1.0	1.4	1.1	0.5
MRSA	MRSA5	1.8	0	0	0.9	0	0
	MRSA6	0.9	0	0	6.9	0	0
	MRSA7	6.3E-2	0.3	0.3	0.1	0.1	0.1
	MRSA8	3.8	1.0	3.4E-6	0.2	0.2	1.9E-4
	MRSA9	0.9	4.2	1.5	1.1	0.4	0.5
	MRSA10	4.4E-2	2.0E-2	5.0E-3	1.8	8.5E-3	6.2E-2

The results are expressed as EOP values. Bold characters indicate host strain.
0 indicates that no plaque was observed with high titer phage preparations.

While the general host ranges remained unchanged, the level of sensitivity as determined by EOP values varied in only a few cases. For example, the three phages had a reduced EOP ($<10^{-4}$) when plated on S. aureus SMQ1292 (ST352, CC97), SMQ1321, and SMQ1322 (ST2270, CC126). Similarly, phage K propagated on S. xylosus had reduced EOP values (10^{-3} to 10^{-5}) when plated on SMQ1286, CMRSA1, CMRSA3 and CMRSA8 strains. The EOP of phage K propagated on S. aureus SA812 dropped also to 10^{-7} and 10^{-6} when plated on CMRSA3 and CMRSA8 strains, respectively (Table 3). A similar phage K behavior was observed in another study [22].

Taken altogether, the host ranges and EOP values were mostly similar when comparing phages amplified on both hosts, indicating that S. xylosus SMQ-121 is a suitable host to produce these myophages. Although the results demonstrate the polyvalent nature of the three phages tested, at least one S. aureus strain was resistant to these polyvalent phages, illustrating the importance of using phage cocktails to cover a wider range of strains. It also suggests that such strain should be used to uncover novel phages.

In conclusion, a new polyvalent staphylococcal phage was characterized. We found that phage Team1, as well as the well-known phages K and phi812, can also be propagated on a non-pathogenic S. xylosus strain and be highly effective in killing a large panel of S. aureus strains obtained from various sources and belonging to different genotypes. This study proposes the use of food-grade bacteria to amplify virulent and polyvalent staphylococcal phages in order to increase the safety of these promising biocontrol agents for both medical and food applications.

SUPPORTING INFORMATION

Table S1. ORF identification, putative function, and comparison of Team1 genome with sequences available in public databases

ORF	Start	End	Length (a.a.)	pI/Mw (KDa)	SD sequence 5'- AGGAGG -3'	Putative function of the deduced protein	Best hit with BLAST	Number of identical a.a./ total number of a.a.(%)	Length (a.a.)	E-value	Accession number (Gen-Bank)
1	496	795	99	4.4/11.6	AGGAGAtgaaatagATG		G1, ORF150	99/99 (100)	99	4.0E-64	YP_241022.1
2	811	996	61	5.7/6.8	AGGAGGGataagtaatcATG		G1, ORF231	61/61 (100)	61	3.0E-30	YP_241023.1
3	1103	1393	96	4.1/11.3	AGGAGAgatataATG		G1, ORF156	96/96 (100)	96	1.0E-60	YP_241024.1
4	1393	1680	95	7.6/10.9	AGGAGAtgtttataATG		G1, ORF158	95/95 (100)	95	2.0E-60	YP_241025.1
5	1680	1973	97	4.3/11.5	AGGAGAgattataATG		G1, ORF154	97/97 (100)	97	3.0E-63	YP_241026.1
6	1977	2234	85	6.1/10.2	AGGAGAtgtaactaATG		G1, ORF175	85/85 (100)	85	3.0E-54	YP_241027.1
7	2312	2551	79	6.0/9.2	AGGGGGgaagttaggATG		G1, ORF183	79/79 (100)	79	1.0E-49	YP_241028.1
8	2562	2909	115	4.5/13.7	AGGAGGggataggATG		K, ORF118	115/115 (100)	115	2.0E-76	YP_024546.1
9	3458	3120	112	4.5/13.5	AAAAGGatttgatttaaATG		G1, ORF128	112/112 (100)	112	1.0E-70	YP_241030.1
10	3769	4077	102	4.7/11.8	TGGAGAtgatttaaATG		G1, ORF145	102/102 (100)	102	1.0E-68	YP_241031.1
11	4283	4570	95	5.2/11.0	AGGATTtgattattATG		GH15, ORF003	92/95 (97)	95	5.0E-61	YP_007002126.1
12	4620	4811	63	9.8/7.7	AGGATATgataatATG		G1, ORF221	63/63 (100)	63	5.0E-37	YP_241033.1
13	5615	5127	162	9.7/19.6	AGGAGGGagtaaaaaaATG	Endo-nucle-ase	G1, ORF085	162/162 (100)	162	7.0E-114	YP_241035.1
14	5783	5941	52	10.4/6.1	AGGGGATtgaacaaaATG		ISP,ORF144	52/52 (100)	52	1.0E-26	CCA65876.1
15	6011	6142	43	10.0/5.1	AGGAAAAtgataaaaATG		G1, ORF297	43/43 (100)	43	5.0E-21	YP_241036.1
16	6309	6632	107	4.8/12.4	AGGAAATtgattataATG		ISP, ORF146	107/107 (100)	107	3.0E-71	CCA65878.1
17	6732	6968	78	4.2/9.1	AGGAAATtgatacaaATG		A5W, ORF194	78/78 (100)	78	9.0E-47	ACB89187.1
18	7048	7518	156	3.6/17.8	AGGATTtgattattATG		A5W, ORF001	156/159 (98)	159	4.0E-104	ACB88992.1
19	7548	7673	41	4.1/4.6	AAGCGGaacaataca-gATG		G1, ORF166	41/89 (46)	89	5.0E-20	YP_241041.1

#				Sequence		ORF	Ratio		E-value	Accession
20	7758	7940	60	5.4/7.4	AGGGAAAtgatgaacATG		SA5, ORF151	60/62 (97)	62	6.0E-33	AFV80806.1
21	8506	8270	78	4.8/9.6	AGGGAGAataagtaatATG		SA5, ORF152	78/240 (32.5)	240	2.0E-49	AFV80807.1
22	8993	8508	161	7.7/19.1	AGGGAGGattaagtATG		G1, ORF088	161/161 (100)	161	4.0E-110	YP_241045.1
23	9413	9006	135	5.2/16.5	AGGGAGGaaaaataATG		G1, ORF109	135/135 (100)	135	3.0E-93	YP_241046.1
24	9844	9413	143	4.5/17.3	AGGGAGGgataaataata-ATG		G1, ORF103	143/143 (100)	143	6.0E-95	YP_241047.1
25	10038	9847	63	9.9/7.9	AGGGAGGactcacaATG		G1, ORF224	63/63 (100)	63	7.0E-36	YP_241048.1
26	10520	10035	161	9.4/18.3	AGGGAGGgaagaagATG		Sb-1, ORF008	161/161 (100)	161	2.0E-109	AEJ79651.1
27	10944	10513	143	4.2/16.7	AGGGAGAgttaaaATG		K, ORF002	143/143 (100)	143	2.0E-97	YP_024433.1
28	11500	10958	180	9.2/21.5	AGTAGGtaattataaaATG		G1, ORF073	180/180 (100)	180	6.0E-125	YP_241051.1
29	12000	11512	162	9.6/19.5	AGGGAGAaatattATG		K, ORF004	162/162 (100)	162	1.0E-116	YP_024435.1
30	12411	12013	132	8.8/16.1	TGGAGGttaagaaATG		K, ORF005	132/132 (100)	132	2.0E-88	YP_024436.1
31	13115	12408	235	4.9/27.7	AGGGAGGgagattagATG		G1, ORF051	235/235 (100)	235	1.0E-170	YP_241054.1
32	13766	13215	183	4.5/21.1	AGGGAGAtttaaatgATG		K, ORF007	183/184 (99)	184	2.0E-128	YP_024438.1
33	14102	13785	105	6.3/11.8	AGGGAGAtaaatttGTG	Major tail protein	G1, ORF138	105/105 (100)	105	5.0E-67	YP_241056.1
34	15636	15088	182	4.5/22.0	GGGGAGTaatatATG		K, ORF008	182/182 (100)	182	1.0E-123	YP_024439.1
35	15858	15640	72	4.4/8.4	AGGGAGAggttagtaaATG		G1, ORF201	72/72 (100)	72	3.0E-43	YP_241058.1
36	16053	15859	64	4.6/7.6	CGGAGGGtctataATG		G1, ORF218	64/64 (100)	64	1.0E-37	YP_241059.1
37	16780	16043	245	6.2/28.7	AGGGAGGgatgttataATG		K, ORF009	245/245 (100)	245	5.0E-174	YP_024440.1
38	16947	16843	34	4.7/4.1	AGGGAGGattgctATG		G1, ORF437	34/34 (100)	34	3.0E-14	YP_241061.1
39	17198	16959	79	4.7/9.3	AGGGATGagtaatATG		A5W, ORF020	79/82 (96)	82	2.0E-50	ACB89011.1
40	17589	17200	129	5.0/15.2	AGGGAGAtgactaattATG		K, ORF010	129/129 (100)	129	8.0E-89	YP_024441.1
41	17861	17688	57	5.1/6.8	AGGGAGGaataaaccctATG		G1, ORF245	57/57 (100)	57	4.0E-34	YP_241064.1
42	18384	17902	160	4.5/18.9	AGGGAGGttggtatATG		K, ORF011	160/160 (100)	160	1.0E-109	YP_024442.1
43	18976	18434	180	4.8/20.4	AGGGAGAggttaagtaATG		K, ORF012	180/180 (100)	180	7.0E-123	YP_024443.1

44	19509	18976	177	4.3/20.7	AGGAGAttaaataaatATG		K, ORF013	177/177 (100)	177	2.0E-123	YP_024444.1
45	19676	19512	54	9.5/6.3	AGGAGTTagagtaactATG		G1, ORF256	54/54 (100)	54	3.0E-29	YP_241068.1
46	19954	19679	91	5.0/10.9	CGGAGGtaggttgtaATG		G1, ORF163	91/91 (100)	91	6.0E-53	YP_241069.1
47	20799	19954	281	9.2/31.7	AGGAGGaaaatcaATG		K, 0RF014	281/281 (100)	281	0	YP_024445.1
48	21929	20811	372	4.7/42.2	AGGAGAAatgataaatATG		G1, ORF024	372/372 (100)	372	0	YP_241071.1
49	22409	22083	108	4.7/13.0	CGGAGAtagtactcGTG		G1, ORF134	108/108 (100)	108	7.0E-73	YP_241072.1
50	22818	22402	138	5.1/16.0	AGGAGGgtattacATG		K, ORF016	138/138 (100)	138	6.0E-95	YP_024447.1
51	23253	22951	100	4.8/11.3	AGGAGAtatagaataATG	DNA binding protein	K, ORF017	100/100 (100)	100	6.0E-64	YP_024448.1
52	23441	23253	62	4.3/7.3	AGGAAGttagaaATG		G1, ORF228	62/62 (100)	62	1.0E-34	YP_241075.1
53	23646	23485	53	4.6/6.4	AGGGAGGttgcctaATG		G1, ORF259	53/53 (100)	53	3.0E-29	YP_241076.1
54	25694	23646	682	6.3/79.8	AGGAGTgattttaaATG		K, ORF018	682/682 (100)	682	0	YP_024449.1
55	26035	25772	87	5.6/10.1	AGGAGGttttataaATG		676Z, ORF038	86/87 (99)	87	8.0E-55	AFN38278.1
56	26225	26052	57	6.6/6.7	AGAGGGtaggtaataTTG		JD007, ORF103	57/57 (100)	57	70E-32	YP_007112812.1
57	26810	26232	192	8.5/21.4	AGGGGAacaagATG		K, ORF019	192/192 (100)	192	1.0E-129	YP_024450.1
58	27429	26803	208	4.6/23.8	AGGAGAtactatagATG		K, ORF020	208/208 (100)	208	3.0E-150	YP_024451.1
59	28318	27422	298	5.4/35.0	AGGAGGataattaATG	DNA ligase	K, ORF021	298/298 (100)	298	0	YP_024452.1
60	28542	28318	74	8.2/8.2	AGGAGGattttttATG		JD007, ORF107	74/74 (100)	74	5.0E-40	YP_007112808.1
61	29351	28611	246	5.2/28.6	AGGGGAactttaaaatATG		K, ORF022	246/246 (100)	246	0	YP_024453.1
62	30017	29403	204	4.0/23.0	AGGAGAaaatatATG		K, ORF023	204/204 (1000)	204	6.0E-144	YP_024454.1
63	30458	30033	141	6.7/15.8	TGGAGGattaaagtATG	Ribonuclease	K, ORF024	141/141 (100)	141	3.00E-94	YP_024455.1
64	30639	30448	63	5.7/7.5	TGGAGGtatttgtttATG		G1, ORF222	63/63 (100)	63	3.00E-37	YP_241086.1
65	31303	30662	213	4.0/24.6	TGGAGGgaagacgATG		K, ORF025	213/213 (100)	213	2.0E-143	YP_024456.1

Start	End	Length	Score	RBS / Start	Description	Phage, ORF	Identity	Length	E-value	Accession
31523	31293	76	8.0/8.8	AGGAGAaataattATG	Transcriptional regulator	G1, ORF187	76/76 (100)	76	1.00E-46	YP_241088.1
31753	31526	75	10.0/9.2	AGGAGGataatgaATG		G1, ORF190	75/75 (100)	75	7.00E-45	YP_241089.1
32555	31863	230	5.0/24.8	AGGAGTtttaaattATG		K, ORF026	230/230 (100)	230	6.0E-167	YP_024457.1
33377	32742	211	9.2/24.8	AGGAGGagatattATG		K, ORF027	211/211 (100)	211	2.0E-151	YP_024458.1
34235	33444	263	8.9/29.3	AGGAGAaagtaATG		K, ORF028	263/263 (100)	263	0	YP_024459.1
34543	34235	102	8.8/12.1	AGGAGGaattacATG		K, ORF029	102/102 (100)	102	7.0E-65	YP_024460.1
35285	34656	209	9.7/23.1	AGTCTGtcaatcaATG	Amidase	G1, ORF060	209/209 (100)	209	5.0E-154	YP_241094.1
36056	35556	166	9.3/19.2	AGGAGAtgttattATG	Endonuclease	K, ORF031	166/166 (100)	166	3.0E-116	YP_024462.1
37019	36216	267	9.5/29.8	AGGAGGaagttaagtaATG	CHAP domain	G1, ORF042	267/267 (100)	267	0.0E+00	YP_241096.1
37522	37019	167	4.1/18.1	AGGTCGgttttaATG	Putative holin	K, ORF033	167/167 (100)	167	7.0E-116	YP_024463.1
37792	37607	61	5.0/7.1	AGGAGAgatacaaATG		G1, ORF233	61/61 (100)	61	1.0E-33	YP_241098.1
39557	39339	72	9.2/8.7	AGGAATtgattaattATG		G1, ORF200	72/72 (100)	72	4.0E-45	YP_241099.1
40244	40035	69	5.7/8.1	AGGAGGgattgtagATG		G1, ORF207	69/69 (100)	69	2.0E-42	YP_241100.1
40589	40257	110	5.0/12.5	AGGTGAttctagTTG		G1, ORF209	110/110 (100)	110	3.0E-69	YP_241101.1
40928	40602	108	5.6/13.0	AGGAGGttaccactTTG	Membrane protein	GH15, ORF077	108/108 (100)	108	6.0E-71	YP_007002200.1
41227	40961	88	9.4/10.1	AGGTGTtttaacaATG		G1, ORF169	88/88 (100)	88	4.0E-50	YP_241102.1
41368	41754	128	9.1/14.8	AGGAGGttaaagtggTTG		G1, ORF168	128/128 (100)	128	7.0E-82	YP_241103.1
41732	42010	92	9.7/10.6	AGGAAGattacaATG		G1, ORF161	92/92 (100)	92	2.0E-60	YP_241104.1
42007	42417	136	4.4/15.6	AGGTGAaatgaTTG		G1, ORF133	136/136 (100)	136	2.0E-91	YP_241105.1

							Twort, ORF151	64/79 (81)	79	4.0E-38	YP_238726.1
85	42432	42629	65	9.6/7.7	AGGAGAaagataaATG	Small subunit of the termi-nase					
86	42923	43894	323	9.1/38.6	AGGTGAaaacagTTG		MSA6, ORF058	323/323 (100)	323	0.	AFN38731.1
87	44035	45582	515	5.8/59.7	AGGTCTtagtgaaATG	Large subunit of the termi-nase	K, ORF035	515/605 (85)	605	0	YP_024465.1
88	45575	46396	273	5.1/30.6	AGAAGGaaataaacaattc-tactttgATG		ISP, ORF002	273/273 (100)	273	0	CCA65732.1
89	46383	46556	57	9.4/6.7	AGGTGAttataGTG		JD007, ORF129	57/57 (100)	57	9.00E-30	YP_007112784.1
90	46553	47032	159	4.8/18.5	AGGAAGgaaataaATG		K, ORF037	159/159 (100)	159	2.0E-109	YP_024467.1
91	47125	48243	372	4.1/40.8	ATGAGGtaatcataaccata-ataacggttATG		K, ORF038	370/397 (93)	397	0	YP_024468.1
92	48320	48670	116	9.3/13.1	CGGTGGgtgaaaactTTG		G1, ORF120	116/166 (100)	116	1.00E-73	YP_240898.1
93	48688	49059	123	5.7/14.5	AGACGGgtgaataggTTG		K, ORF040	123/123 (100)	123	9.00E-83	YP_024470.1
94	49063	50754	563	6.1/64.1	AGGTGActagtaaTTG	Portal protein	K, ORF041	563/563 (100)	563	0	YP_024471.1
95	50948	51721	257	4.9/28.6	TGGAGGtgtaga-cacctTTG	Prohead prote-ase	K, ORF042	257/257 (100)	257	0	YP_024472.1
96	51740	52696	318	4.4/36.0	AGGAGAatacattctATG		G1, ORF029	318/318 (100)	318	0	YP_240902.1
97	52812	54203	463	5.1/51.2	AGGTGAtaaatttatATG	Major capsid protein	K, ORF044	463/463 (100)	463	0	YP_024474.1
98	54295	54591	98	9.4/11.3	AGGGATttaataaatATG		G1, ORF151	98/98 (100)	98	1.00E-59	YP_240904.1
99	54604	55512	302	5.1/34.2	AGGGTGaattaaATG		K, ORF045	302/302 (100)	302	0	YP_024475.1
100	55526	56404	292	5.6/33.8	AGGGAGGgttagaaaATG	Capsid protein	K, ORF046	292/292 (100)	292	0	YP_024476.1

#	Start	End	Length	Ratio	SD sequence	Protein	ORF	Identity	Length	E-value	Accession
101	56404	57024	206	10.3/23.8	CGGAGGtgcatttaaata-ATG		K, ORF047	206/206 (100)	206	2.0E-148	YP_024477.1
102	57043	57879	278	4.7/31.8	AGGAGGgttagtattaaATG		K, ORF048	278/278 (100)	278	0	YP_024478.1
103	57881	58096	71	8.0/8.3	TGGAAGtaggATG		G1, ORF202	71/71 (100)	71	2.0E-45	YP_240909.1
104	58123	59886	587	4.9/64.5	AGGAGAattaaatATG	Tail sheath protein	K, ORF049	586/587 (99)	587	0	YP_024479.1
105	59959	60387	142	5.4/16.0	AGGAGAgtgaataca-gATG	Structural protein	K, ORF050	142/142 (100)	142	1.0E-99	YP_024480.1
106	60484	60624	46	10.7/5.4	AGGAGAtgtactagATG		G1, ORF293	43/43 (100)	43	4.0E-23	YP_240912.1
107	60667	61125	152	9.6/18.1	AGGAGAtgggtatATG		K, ORF051	152/152 (100)	152	1.0E-107	YP_024481.1
108	61138	61332	64	9.5/7.1	AGGAGGtattataATG		G1, ORF215	64/64 (100)	64	2.0E-34	YP_240914.1
109	61414	61725	103	5.9/12.3	AGGAGAaagatataaaATG		K, ORF052	103/103 (100)	103	8.0E-66	YP_024482.1
110	61928	62314	128	4.7/15.3	AGATGTTagtaaaATG		K, ORF053	128/152 (84)	152	2.0E-88	YP_024483.1
111	62358	62894	178	4.2/21.0	AGAACGattgggcg-gtATG	RNA poly-merase	K, ORF054	178/178 (100)	178	1.0E-125	YP_024484.1
112	62950	63093	47	4.0/5.3	TGGATCggtgaataatgATG		GH15, ORF107	46/1352 (0.03)	1352	2.0E-21	YP_007002230.1
113	63132	67004	1290	9.2/137.0	AGAACGcaATG	Tail measure protein	G1, ORF001	1290/1352 (95)	1352	0	YP_240918.1
114	67083	69509	808	6.3/91.2	AGGAAGtatgtgtatATG	Tail endoly-sin	K, ORF056	808/808 (100)	808	0	YP_024486.1
115	69523	70410	295	4.5/34.6	AGGAGGatagtctATG		K, ORF057	295/295 (100)	295	0	YP_024487.1
116	70410	72956	848	4.8/96.1	AAGAGGgtatatttaATG		G1, ORF004	848/848 (100)	848	0	YP_240923.1
117	73062	73853	263	7.8/29.3	AGGAGGaggataaATG		K, ORF059	263/263 (100)	263	0	YP_024489.1
118	73853	74377	174	4.5/20.0	TGGAGGtgtatctagcta-ATG		G1, ORF078	174/174 (100)	174	3.0E-121	YP_240925.1

119	74377	75081	234	4.6/26.6	AGGAGGctaacgtataATG	Base-plate protein	K, ORF061	234/234 (100)	234	5.0E-171	YP_024491.1
120	75096	76142	348	4.7/39.2	AGGATTaaattATG	Tail protein	K, ORF062	348/348	348	0	YP_024492.1
121	76163	79222	1019	5.0/116.3	AGGTGAaacttaagtcGTG		G1, ORF003	1019/1019 (100)	1019	0	YP_240928.1
122	79333	79854	173	5.3/19.2	AGGAATtaaaaaatATG	Structural protein	K, ORF064	173/173 (100)	173	3.0E-122	YP_024494.1
123	79875	83333	1152	5.1/129.1	GGGAGAAtaattctaaATG	Adsorption associated protein	K, ORF065	1152/1152 (100)	1152	0	YP_024495.1
124	83382	83540	52	8.0/6.2	AGGAGAgatttatATG		G1, ORF262	52/52 (100)	52	8.0E-27	YP_240931.1
125	83541	85463	640	6.3/72.6	AGGAGAaaaatagATG		G1, ORF009	639/640 (99)	640	0	YP_240932.1
126	85486	85860	124	4.7/14.6	AGGTGGaataaaaaac-tATG		K, ORF067	124/124 (100)	124	10E-83	YP_024497.1
127	85867	87243	458	5.9/50.4	AGGGGTaattataaATG		G1, ORF017	458/458 (100)	458	0	YP_240934.1
128	87334	89082	582	5.6/67.2	AGGAGATtaaaATG	Helicase	K, ORF069	582/582 (100)	582	0	YP_024499.1
129	89094	90707	537	8.0/63.2	AGGAGGgtaagagATG		K ,ORF070	537/537 (100)	537	0	YP_024500.1
130	90700	92142	480	5.5/54.6	AAGAGGttaataactATG	DNA helicase	K, ORF071	480/480 (100)	480	0	YP_024501.1
131	92221	93258	345	4.7/40.1	AGGAGAgattaataATG	Exonuclease	K, ORF072	345/345 (100)	345	0	YP_024502.1
132	93258	93635	125	5.2/14.9	AGGAGGGtttataATG		K, ORF073	125/125 (100)	125	2.0E-85	YP_024503.1
133	93635	95554	639	5.1/73.4	AGGGGGAtaagtaATG	Exonuclease	K, ORF074	639/639 (100)	639	0	YP_024504.1
134	95554	96150	198	6.3/23.2	TGGAGGaaaaataATG		K, ORF075	198/198 (100)	198	5.0E-142	YP_024505.1
135	96165	97232	355	8.5/40.9	AGGAAGgatattATG	DNA primase	K, ORF076	355/355 (100)	355	0	YP_024506.1

136	97298	97636	112	4.2/13.0	AGGAGAaaaaataATG		G1, ORF127	112/112 (100)	112	1.0E-71	YP_240943.1
137	97636	98088	150	4.8/17.1	AGGAGAacaagaataATG		Sb-1, ORF120	150/150 (100)	150	4.0E-100	AEJ79755.1
138	98075	98683	202	5.5/23.6	AGTTGaagaaaaATG	Re-solvase	G1, ORF064	202/202 (100)	202	1.0E-147	YP_240945.1
139	98661	99092	143	9.8/16.2	ATGAATtgattaATG		A5W, ORF109	143/143 (100)	143	4.0E-97	ACB89102.1
140	99107	101221	704	5.6/80.1	AGGATGaagagatatATG		G1, ORF006	702/704 (99)	704	0	YP_240947.1
141	101235	102284	349	4.7/40.5	AGGAAAtagatATG		K, ORF081	349/349 (100)	349	0	YP_024511.1
142	102302	102631	109	4.5/12.4	AGGAGAaaagatattATG		K, ORF082	109/109 (100)	109	2.0E-72	YP_024512.1
143	102615	102935	106	4.8/12.1	AGGATGattcagaagATG		K, ORF083	106/106 (100)	106	6.0E-69	YP_024513.1
144	103178	103738	186	6.5/22.1	AGAAGAtatatcATG		K, ORF084	186/198 (94)	198	1.0E-131	YP_024514.1
145	103748	104053	101	5.8/11.9	AGGTGAttatATG	Integration host factor	K, ORF085	101/101 (100)	101	4.0E-66	YP_024515.1
146	104129	105001	290	5.9/33.2	AGGAGAggaattaaATG	DNA polymerase	G1, ORF035	290/290 (100)	290	0	YP_240953.1
147	105167	105679	170	9.7/20.3	AGGTGAgacataGTG		G1, ORF081	170/170 (100)	170	3.0E-118	YP_240954.1
148	105815	107158	447	5.3/52.8	ATTAAGttcttaATG	DNA polymerase	K, ORF86	445/1072 (41)	1072	0	YP_024516.1
149	107426	108133	235	9.6/27.5	AGGAGAgttatATG	HNH endonuclease	K, ORF089	235/269 (87)	269	4.0E-169	YP_024518.1
150	108367	109227	286	4.9/32.9	AGGGTCcgactatagc-gccctagagATG	DNA polymerase	G1, ORF037	286/286 (100)	286	0	YP_240958.1
151	109296	109538	80	4.2/9.1	AGGAGGaaaaagaGTG		G1, ORF181	80/80 (100)	80	7.0E-50	YP_240959.1
152	109555	110037	160	5.5/18.9	AGGAGGgaactgataaATG		G1, ORF091	160/160 (100)	160	1.0E-114	YP_024519.1
153	110124	111395	423	4.6/46.9	AGGAGAaaagataattATG		K, ORF092	423/423 (100)	423	0	YP_024520.1

154	111455	111679	74	6.1/7.9	AGGAGGatattaaATG	Recombinase	K, ORF093	71/418 (17)	418	1.0E-39	YP_024521.1
155	112024	112992	322	9.3/38.3	TGGAGAgtagtataatATG	Endonuclease	SA11, ORF083	282/322 (88)	322	0	YP_007005558.1
156	113140	114087	315	5.1/35.7	AGCAGAtaacaatcgtATG	Recombinase	K, ORF093	315/418 (75)	418	0	YP_024521.1
157	114091	114444	117	5.1/13.4	TGGAGAttaagcttATG		G1, ORF121	117/117 (100)	117	5.0E-79	YP_240963.1
158	114431	115093	220	5.3/26.6	AGGGGGctaatccttATG		K, ORF094	220/220 (100)	220	9.0E-156	YP_024522.1
159	115221	115853	210	4.7/23.2	TGGAGAttaagcttATG		K, ORF095	210/210 (100)	210	2.0E-149	YP_024523.1
160	115876	116388	170	4.3/17.9	AGGATGgttaaaTTG	Major tail protein	G1, ORF080	170/170 (100)	170	1.0E-113	YP_240966.1
161	116403	116630	75	4.4/7.8	AGGAGGacaata-aagaATG	Major tail protein	G1, ORF189	75/75 (100)	75	8.0E-45	YP_240967.1
162	116726	116986	86	5.6/10.3	AGGAGGgggaagtaATG		G1, ORF174	86/86 (100)	86	1.0E-54	YP_240968.1
163	116990	117745	251	4.4/29.2	AGTAGGtgagtagggtATG		K, ORF097	251/251 (100)	251	6.0E-180	YP_024525.1
164	117738	118988	416	5.7/47.5	AGGAAGcagttATG	DNA poly-merase	K, ORF098	416/416 (100)	416	0	YP_024526.1
165	119002	119370	122	5.6/14.0	AGGATGtgttaataaATG		K, ORF099	122/122 (100)	122	2.0E-80	YP_024527.1
166	119357	119668	103	4.6/12.0	AGGAGGataataATG		K, ORF100	103/103 (100)	103	6.0E-68	YP_024528.1
167	119732	120268	178	6.9/20.8	AGGTGActtaaaaATG		G1, ORF075	178/178 (100)	178	5.0E-127	YP_240973.1
168	120261	121028	255	9.6/30.1	AGGAGAtagttaaATG		G1, ORF045	255/255 (100)	255	0	YP_240974.1
169	121006	121398	130	9.6/15.1	AAGAGTaattaATG		SA5, ORF080	130/130 (100)	1630	2.0E-88	AFV80741.1
170	121451	122314	287	5.5/32.4	AGGAAAgataaataATG		G1, ORF036	287/287 (100)	287	0	YP_240976.1
171	122686	123417	243	5.2/28.4	AGGAGAgctatataATG		K, ORF103	243/243 (100)	243	1.0E-172	YP_024531.1
172	123435	123893	152	4.8/17.8	GTGAGGtatagagtaATG		G1, ORF094	152/152 (100)	152	5.0E-105	YP_240978.1
173	123958	124401	147	6.0/17.5	AGGAGCtaacaattATG		K, ORF105	147/147 (100)	147	2.0E-98	YP_024533.1

174	124418	125122	234	4.6/27.4	AAGAGGttataatATG		K, ORF106	234/234 (100)	234	5.0E-168	YP_024534.1
175	125184	125582	132	8.9/15.4	TAGAGGtgttaattATG		K, ORF107	132/132 (100)	132	5.0E-90	YP_024535.1
176	125729	125971	80	9.3/9.4	AGGAGAgattaactATG		G1, ORF182	80/80 (100)	80	3.0E-48	YP_240982.1
177	125976	126140	54	6.2/6.3	AGGAGAtaggacaATG		G1, ORF252	54/54 (100)	54	2.0E-29	YP_240983.1
178	126127	126306	59	9.3/7.1	ATTAGAtttctattaggTTG		A5W, ORF146	59/59 (100)	59	2.0E-33	ACB89139.1
179	126342	126518	58	4.6/7.0	AGGAGGagagatattATG		G1, ORF240	58/58 (100)	58	2.0E-32	YP_240984.1
180	126924	127043	39	8.0/4.9	TGGAAAaaccaaTTG		G1, ORF076	38/177 (22)	177	2.0E-16	YP_240985.1
181	127058	127306	82	4.7/9.1	AGGTGGgaaagcaATG		ISP, ORF92	82/82 (100)	82	7.0E-45	CCA65823.1
182	127318	127494	58	9.9/7.0	AGGAGAtttactATG		G1, ORF241	58/58 (100)	58	6.0E-31	YP_240986.1
183	127487	127783	98	6.6/11.3	AGGAGAaaaagaaATG		G1, ORF152	98/98 (100)	98	3.0E-63	YP_240987.1
184	127831	128013	60	8.2/7.2	AAGAGGgaatgattattATG		G1, ORF219	59/60 (98)	60	2.0E-32	YP_240988.1
185	128026	128394	122	4.4/14.2	AGGAGGttgtatagATG		G1, ORF119	122/122 (100)	122	1.0E-81	YP_240989.1
186	128407	128754	115	4.6/13.0	AGGAGGaaatagaATG		G1, ORF124	115/115 (100)	115	2.0E-75	YP_240990.1
187	128754	129032	92	4.3/10.2	TGGAAAgactggaactgggattaATG		G1, ORF162	92/92 (100)	92	5.0E-55	YP_240991.1
188	129102	129407	101	9.2/12.1	AGGAGAtgatactATG		G1, ORF140	101/101 (100)	101	7.0E-67	YP_240992.1
189	129422	129772	116	10.0/13.7	AGGAGGtggaaagATG		G1, ORF122	116/116 (100)	116	4.0E-76	YP_240993.1
190	129772	130374	200	9.7/23.4	AGGAGGaaaaataATG		G1, ORF065	200/200 (100)	200	2.0E-144	YP_240994.1
191	130388	130567	59	9.0/7.3	AGGTGGttatataaATG		G1, ORF237	59/59 (100)	59	2.0E-34	YP_240995.1
192	130793	131194	133	4.9/15.0	AGGAGTtggttgtaATG		G1, ORF107	133/133 (100)	133	2.0E-86	YP_240996.1
193	131196	131456	86	4.4/10.1	AGGAGGgaaattaaaATG		G1, ORF173	86/86 (100)	86	4.0E-53	YP_240997.1
194	131508	131795	95	9.2/10.5	AGGAGGgaaattaatATG		G1, ORF157	95/95 (100)	95	5.0E-59	YP_240999.1
195	131806	131922	38	4.8/4.6	AGGAGGgaattctATG	Tran-scription factor	G1, ORF362	38/38 (100)	38	8.0E-15	YP_241000.1
196	131912	132175	87	10.1/9.9	AGGAGGaaaaacaaaa-gATG		G1, ORF170	87/87 (100)	87	7.0E-53	YP_241001.1

197	132252	132431	59	10.1/6.4	AGGAGGttaatttATG	G1, ORF236	59/59 (100)	59	3.0E-29	YP_241002.1
198	132446	132709	87	4.8/10.3	AGGAGGgaactataaATG	G1, ORF171	87/87 (100)	87	1.0E-55	YP_241003.1
199	132712	133029	105	5.0/12.0	AGGAGGgaaattaattATG	G1, ORF137	105/105 (100)	105	7.0E-68	YP_241004.1
200	133030	133128	32	8.2/3.6	AGGAGGgaaaataaGTG	G1, ORF055	32/226 (14)	226	9.0E-13	YP_241006.1
201	133396	133710	104	5.1/11.6	ATGAATattattcaATG	G1, ORF055	104/226 (46)	226	4.0E-66	YP_241006.1
202	133799	133957	52	5.8/5.7	AGGAGGtaatataATG	G1, ORF263	52/52 (100)	52	3.0E-23	YP_241007.1
203	133992	134192	66	5.1/7.6	AGGAGGAtggtaaatATG	G1, ORF211	66/66 (100)	66	4.0E-41	YP_241008.1
204	134193	134483	96	8.9/11.1	CGGAGGgggttataaATG	G1, ORF155	96/96 (100)	155	5.0E-59	YP_241009.1
205	134575	134883	102	5.5/12.0	AGGAGAtgaacaattATG	K, ORF109	102/102 (100)	102	2.0E-63	YP_024537.1
206	134880	135788	302	5.1/35.2	AGGAGCtatcaacaaaATG	K, ORF110	302/302 (100)	302	0	YP_024538.1
207	135806	137275	489	5.3/56.1	AGGAGGAaaataattATG	K, ORF111	489/489 (100)	489	0	YP_024539.1
208	137354	137599	81	8.8/10.0	AGGAGTgaaagaaATG	G1, ORF178	81/81 (100)	81	2.0E-51	YP_241013.1
209	137619	138011	130	5.1/15.4	AGGAGAgaaataaaATG	K, ORF113	130/130 (100)	130	2.0E-86	YP_241014.1
210	138013	138234	73	4.5/8.9	AGGAGAgaaatagcATG	G1, ORF194	72/73 (99)	73	2.0E-43	YP_241015.1
211	138300	138611	103	5.1/11.6	AGGAGGAtgaaatATG	G1, ORF142	103/103 (100)	103	1.0E-65	YP_241016.1
212	138614	139123	169	9.3/20.3	AGGAGGActaactATG	G1, ORF082	169/169 (100)	169	6.0E-118	YP_241017.1
213	139125	139454	109	5.3/12.6	AGGAAGtgtttaagta-atATG	G1, ORF131	109/109 (100)	109	3.0E-73	YP_241018.1
214	139460	139654	64	7.8/7.8	GGGAGGTaattataATG	A5W, ORF179	64/64 (100)	64	1.0E-36	ACB89172.1
215	139678	139992	104	4.1/12.0	AGGAGGAgtaaactat-tATG	G1, ORF139	104/104 (100)	104	2.0E-66	YP_241019.1
216	139986	140174	62	4.3/7.3	AGCAGGgttagattactta-aatattaaATG	A5W, ORF181	62/62 (100)	62	3.0E-35	ACB89174.1
217	140211	140312	33	5.1/3.7	AGGAGGaaacatATG	G1, ORF445	33/33 (100)	33	2.0E-13	YP_241021.1

ACKNOWLEDGEMENT

We thank Barbara-Ann Conway for editorial assistance. We also would like to thank the Ministère de l'Agriculture, des Pêcheries et de l' Alimentation du Québec (MAPAQ), the Canadian Bovine Mastitis Research Network and the National Microbiology Laboratory of the Public Health Agency of Canada for providing the *S. aureus* strains. L.H. was the recipient of a scholarship from the Canadian International Development Agency. J.C. holds the Canada Research Chair in Medical Genomics. S.M. holds the Canada Research Chair in Bacteriophages.

AUTHOR CONTRIBUTIONS

Conceived and designed the experiments: SM DSG JC SL LEH. Performed the experiments: LEH NBA PLP. Analyzed the data: LEH NBA PLP JD RK SL JC DSG SM. Contributed reagents/materials/analysis tools: LEH NBA PLP RK JC DSG SM. Contributed to the writing of the manuscript: LEH NBA PLP JD RK SL JC DSG SM.

REFERENCES

1.   Lowy FD (1998) *Staphylococcus aureus* infections. N Engl J Med 339: 520–532. doi: 10.1056/nejm199808203390806

2.   Le Loir Y, Baron F, Gautier M (2003) *Staphylococcus aureus* and food poisoning. Genet Mol Res 2: 63–76.

3.   Golding GR, Campbell JL, Spreitzer DJ, Veyhl J, Surynicz K, et al. (2008) A preliminary guideline for the assignment of methicillin-resistant *Staphylococcus aureus* to a Canadian pulsed-field gel electrophoresis epidemic type using *spa* typing. Can J Infect Dis Med Microbiol 19: 273–281.

4.   Endo Y, Yamada T, Matsunaga K, Hayakawa Y, Kaidoh T, et al. (2003) Phage conversion of exfoliative toxin A in *Staphylococcus aureus* isolated from cows with mastitis. Vet Microbiol 96: 81–90. doi: 10.1016/s0378-1135(03)00205-0

5.   Kaneko J, Kimura T, Kawakami Y, Tomita T, Kamio Y (1997) Panton-valentine leukocidin genes in a phage-like particle isolated from mitomycin C-treated*Staphylococcus aureus* V8 (ATCC 49775). Biosci Biotechnol Biochem 61: 1960–1962. doi: 10.1271/bbb.61.1960

6.   Tinsley CR, Bille E, Nassif X (2006) Bacteriophages and pathogenicity: more than just providing a toxin? Microbes Infect 8: 1365–1371. doi: 10.1016/j.micinf.2005.12.013

7. El Haddad L, Moineau S (2013) Characterization of a novel Panton-Valentine leukocidin (PVL)-encoding staphylococcal phage and its naturally PVL-lacking variant. Appl Environ Microbiol 79: 2828–2832. doi: 10.1128/aem.03852-12

8. Deghorain M, Van Melderen L (2012) The Staphylococci phages family: an overview. Viruses 4: 3316–3335. doi: 10.3390/v4123316

9. Fortier LC, Sekulovic O (2013) Importance of prophages to evolution and virulence of bacterial pathogens. Virulence 4: 354–365. doi: 10.4161/viru.24498

10. BIOHAZ (2012) EFSA Panel on Biological Hazards. Scientific opinion on the evaluation of the safety and efficacy of Listex™ P100 for the removal of Listeria monocytogenes surface contamination of raw fish. EFSA J 10: 2615–2658.

11. Carlton RM, Noordman WH, Biswas B, de Meester ED, Loessner MJ (2005) Bacteriophage P100 for control of *Listeria monocytogenes* in foods: genome sequence, bioinformatic analyses, oral toxicity study, and application. Regul Toxicol Pharmacol 43: 301–312. doi: 10.1016/j.yrtph.2005.08.005

12. Chibeu A, Agius L, Gao A, Sabour PM, Kropinski AM, et al. (2013) Efficacy of bacteriophage Listex™ P100 combined with chemical antimicrobials in reducing Listeria monocytogenes in cooked turkey and roast beef. Int J Food Microbiol 167: 208–214. doi: 10.1016/j.ijfoodmicro.2013.08.018

13. Santos SB, Fernandes E, Carvalho CM, Sillankorva S, Krylov VN, et al. (2010) Selection and characterization of a multivalent *Salmonella* phage and its production in a nonpathogenic *Escherichia coli* strain. Appl Environ Microbiol 76: 7338–7342. doi: 10.1128/aem.00922-10

14. Cui Z, Song Z, Wang Y, Zeng L, Shen W, et al. (2012) Complete genome sequence of wide-host-range *Staphylococcus aureus* phage JD007. J Virol 86: 13880–13881. doi: 10.1128/jvi.02728-12

15. Garcia P, Martinez B, Obeso JM, Lavigne R, Lurz R, et al. (2009) Functional genomic analysis of two *Staphylococcus aureus* phages isolated from the dairy environment. Appl Environ Microbiol 75: 7663–7673. doi: 10.1128/aem.01864-09

16. Kvachadze L, Balarjishvili N, Meskhi T, Tevdoradze E, Skhirtladze N, et al. (2011) Evaluation of lytic activity of staphylococcal bacteriophage Sb-1 against freshly isolated clinical pathogens. Microb Biotechnol 4: 643–650. doi: 10.1111/j.1751-7915.2011.00259.x

17. Gu J, Liu X, Lu R, Li Y, Song J, et al. (2012) Complete genome sequence of *Staphylococcus aureus* bacteriophage GH15. J Virol 86: 8914–8915. doi: 10.1128/jvi.01313-12

18. Vandersteegen K, Mattheus W, Ceyssens P-J, Bilocq F, De Vos D, et al. (2011) Microbiological and molecular assessment of bacteriophage ISP for the control of *Staphylococcus aureus*. PLoS ONE 6: e24418. doi: 10.1371/journal.pone.0024418

19. Capparelli R, Parlato M, Borriello G, Salvatore P, Iannelli D (2007) Experimental phage therapy against *Staphylococcus aureus* in mice. Antimicrob Agents Chemother 51: 2765–2573. doi: 10.1128/aac.01513-06

20. Han JE, Kim JH, Hwang SY, Choresca CH Jr, Shin SP, et al. (2013) Isolation and characterization of a *Myoviridae* bacteriophage against *Staphylococcus aureus* isolated from dairy cows with mastitis. Res Vet Sci 95: 758–763. doi: 10.1016/j.rvsc.2013.06.001

21. Hsieh SE, Lo HH, Chen ST, Lee MC, Tseng YH (2011) Wide host range and strong lytic activity of *Staphylococcus aureus* lytic phage Stau2. Appl Environ Microbiol 77: 756–761. doi: 10.1128/aem.01848-10

22. O›Flaherty S, Ross RP, Meaney W, Fitzgerald GF, Elbreki MF, et al. (2005) Potential of the polyvalent anti-*Staphylococcus* bacteriophage K for control of antibiotic-resistant staphylococci from hospitals. Appl Environ Microbiol 71: 1836–1842. doi: 10.1128/aem.71.4.1836-1842.2005

23. Pantucek R, Rosypalova A, Doskar J, Kailerova J, Ruzickova V, et al. (1998) The polyvalent staphylococcal phage phi812: its host-range mutants and related phages. Virology 246: 241–252. doi: 10.1006/viro.1998.9203

24. Krzywinski M, Schein J, Birol I, Connors J, Gascoyne R, et al. (2009) Circos: an information aesthetic for comparative genomics. Genome Res 19: 1639–1645. doi: 10.1101/gr.092759.109

25. Dordet-Frisoni E, Dorchies G, De Araujo C, Talon R, Leroy S (2007) Genomic diversity in *Staphylococcus xylosus*. Appl Environ Microbiol 73: 7199–7209. doi: 10.1128/aem.01629-07

26. Barriere C, Centeno D, Lebert A, Leroy-Setrin S, Berdague JL, et al. (2001) Roles of superoxide dismutase and catalase of *Staphylococcus xylosus* in the inhibition of linoleic acid oxidation. FEMS Microbiol Lett 201: 181–185. doi: 10.1111/j.1574-6968.2001.tb10754.x

27. Benito MJ, Serradilla MJ, Martin A, Aranda E, Hernandez A, et al. (2008) Differentiation of Staphylococci from Iberian dry fermented sausages

by protein fingerprinting. Food Microbiol 25: 676–82. doi: 10.1016/j.fm.2008.03.007

28. Enright MC, Day NP, Davies CE, Peacock SJ, Spratt BG (2000) Multilocus sequence typing for characterization of methicillin-resistant and methicillin-susceptible clones of*Staphylococcus aureus*. J Clin Microbiol 38: 1008–1015.

29. Maidak BL, Olsen GJ, Larsen N, Overbeek R, McCaughey MJ, et al. (1996) The Ribosomal Database Project (RDP). Nucl Acids Res 24: 82–85. doi: 10.1093/nar/24.1.82

30. Feil EJ, Li BC, Aanensen DM, Hanage WP, Spratt BG (2004) eBURST: inferring patterns of evolutionary descent among clusters of related bacterial genotypes from multilocus sequence typing data. J Bacteriol 186: 1518–1530. doi: 10.1128/jb.186.5.1518-1530.2004

31. Jikia D, Chkhaidze N, Imedashvili E, Mgaloblishvili I, Tsitlanadze G, et al. (2005) The use of a novel biodegradable preparation capable of the sustained release of bacteriophages and ciprofloxacin, in the complex treatment of multidrug-resistant*Staphylococcus aureus*-infected local radiation injuries caused by exposure to Sr90. Clin Exp Dermatol 30: 23–26. doi: 10.1111/j.1365-2230.2004.01600.x

32. .Markoishvili K, Tsitlanadze G, Katsarava R, Morris JG Jr, Sulakvelidze A (2002) A novel sustained-release matrix based on biodegradable poly(ester amide)s and impregnated with bacteriophages and an antibiotic shows promise in management of infected venous stasis ulcers and other poorly healing wounds. Int J Dermatol 41: 453–458. doi: 10.1046/j.1365-4362.2002.01451.x

33. Moineau S, Durmaz E, Pandian S, Klaenhammer TR (1993) Differentiation of two abortive mechanisms by using monoclonal antibodies directed toward lactococcal bacteriophage capsid proteins. Appl Environ Microbiol 59: 208–212.

34. Deveau H, Van Calsteren MR, Moineau S (2002) Effect of exopolysaccharides on phage-host interactions in *Lactococcus lactis*. Appl Environ Microbiol 68: 4364–4369. doi: 10.1128/aem.68.9.4364-4369.2002

35. Bonfield JK, Smith KF, Staden R (1995) A new DNA sequence assembly program. Nucleic Acids Res 23: 4992–4999. doi: 10.1093/nar/23.24.4992

36. Hall TA (1999) BioEdit: a user-friendly biological sequence alignment editor and analysis program for Windows 95/98/NT. Nucleic Acids Symposium Series 41: 95–98.

37. Rombel IT, Sykes KF, Rayner S, Johnston SA (2002) ORF-FINDER: a vector for high-throughput gene identification. Gene 282: 33–41. doi: 10.1016/s0378-1119(01)00819-8

38. Besemer J, Lomsadze A, Borodovsky M (2001) GeneMarkS: a self-training method for prediction of gene starts in microbial genomes. Implications for finding sequence motifs in regulatory regions. Nucleic Acids Res 29: 2607–2618. doi: 10.1093/nar/29.12.2607

39. Kwan T, Liu J, DuBow M, Gros P, Pelletier J (2005) The complete genomes and proteomes of 27 *Staphylococcus aureus* bacteriophages. Proc Natl Acad Sci 102: 5174–5179. doi: 10.1073/pnas.0501140102

40. Conesa A, Gotz S, Garcia-Gomez JM, Terol J, Talon M, et al. (2005) Blast2GO: a universal tool for annotation, visualization and analysis in functional genomics research. Bioinformatics 21: 3674–3676. doi: 10.1093/bioinformatics/bti610

41. Lowe TM, Eddy SR (1997) tRNAscan-SE: A Program for Improved Detection of Transfer RNA Genes in Genomic Sequence. Nucleic Acids Res 25: 955–964. doi: 10.1093/nar/25.5.955

42. Laslett D, Canback B (2004) ARAGORN, a program to detect tRNA genes and tmRNA genes in nucleotide sequences. Nucleic Acids Res 32: 11–16. doi: 10.1093/nar/gkh152

43. Dlawer A, Hiramatsu K (2004) *Staphylococcus aureus*: molecular and clinical aspects. Japan: Horwood Publishing Limited. 267 p.

44. Kloos WE, Schleifer KH (1975) Isolation and characterization of staphylococci from human skin II. Descriptions of four new species: *Staphylococcus warneri,Staphylococcus capitis*, *Staphylococcus hominis*, and *Staphylococcus simulans*. Int J Syst Bacteriol 25: 62–79. doi: 10.1099/00207713-25-1-62

45. Lobocka M, Hejnowicz MS, Dabrowski K, Gozdek A, Kosakowski J, et al. (2012) Genomics of staphylococcal Twort-like phages - potential therapeutics of the post-antibiotic era. Adv Virus Res 83: 143–216. doi: 10.1016/b978-0-12-394438-2.00005-0

46. Vandersteegen K, Kropinski AM, Nash JH, Noben JP, Hermans K, et al. (2013) Romulus and Remus, two phage isolates representing a distinct clade within the *Twortlikevirus* genus, display suitable properties for phage therapy applications. J Virol 87: 3237–3247. doi: 10.1128/jvi.02763-12

47. Stothard P (2000) The sequence manipulation suite: JavaScript programs for analyzing and formatting protein and DNA sequences. Biotechniques 28: 1102–1104.

48. Marcó MB, Garneau JE, Tremblay D, Quiberoni A, Moineau S (2012) Characterization of Two Virulent Phages of *Lactobacillus plantarum*. Appl Environ Microbiol 78: 8719–8734. doi: 10.1128/aem.02565-12

49. Barkema HW, Green MJ, Bradley AJ, Zadoks RN (2009) Invited review: The role of contagious disease in udder health. J Dairy Sci 92: 4717–4729. doi: 10.3168/jds.2009-2347

50. Hata E, Katsuda K, Kobayashi H, Uchida I, Tanaka K, et al. (2010) Genetic variation among *Staphylococcus aureus* strains from bovine milk and their relevance to methicillin-resistant isolates from humans. J Clin Microbiol 48: 2130–2139. doi: 10.1128/jcm.01940-09

51. Wolf C, Kusch H, Monecke S, Albrecht D, Holtfreter S, et al. (2011) Genomic and proteomic characterization of *Staphylococcus aureus* mastitis isolates of bovine origin. Proteomics 11: 2491–2502. doi: 10.1002/pmic.201000698

52. Rabello RF, Moreira BM, Lopes RM, Teixeira LM, Riley LW, et al. (2007) Multilocus sequence typing of *Staphylococcus aureus* isolates recovered from cows with mastitis in Brazilian dairy herds. J Med Microbiol 56: 1505–1511. doi: 10.1099/jmm.0.47357-0

53. Montgomery CP, Boyle-Vavra S, Adem PV, Lee JC, Husain AN, et al. (2008) Comparison of virulence in community-associated methicillin-resistant *Staphylococcus aureus* pulsotypes USA300 and USA400 in a rat model of pneumonia. J Infect Dis 198: 561–570. doi: 10.1086/590157

54. Sakwinska O, Giddey M, Moreillon M, Morisset D, Waldvogel A, et al. (2011)*Staphylococcus aureus* host range and human-bovine host shift. Appl Environ Microbiol 77: 5908–5915. doi: 10.1128/aem.00238-11

Chapter 4

A NEW STRATEGY FOR THE PREPARATION OF N-AMINOPIPERIDINE USING HYDROXYLAMINE-O-SULFONIC ACID: SYNTHESIS, KINETIC MODELLING, PHASE EQUILIBRIA, EXTRACTION AND PROCESSES

E. Labarthe, A. J. Bougrine, Véronique Pasquet, H. Delalu

Laboratoire Hydrazines et Composés Energétiques Polyazotés, UMR CNRS-CNES-SME-Groupe Safran, Université Claude Bernard Lyon 1, Villeurbanne, France

ABSTRACT

A new strategy for the synthesis of N-aminopiperidine (NAPP) was developed using hydroxylamine-O-sulfonic acid (HOSA). A systematic study of NAPP formation and degradation reactions was carried out in diluted medium, in order to identify products and to establish a kinetic modelling. Principal parameters have been defined, in particular, that obtaining high yields (>90%) requires non stoichiometric conditions. The extraction and purification processes were also studied. NAPP isolation and piperidine recycling were optimized after the establishment of the various solid-liquidliquid and liquid-vapour implied phase diagrams. At least, a calorimetric study of solvatation and reaction enthalpies was undertaken in order to estimate reactor heating temperature in the case of anhydrous synthesis. The combination of our kinetic, thermodynamic and calorimetric data allows the establishment of two process schemes: one using pure piperidine, the other, a 66 w% titrating azeotropic solution in piperidine.

INTRODUCTION

N-aminopiperidine (NAPP) is an exocyclic hydrazine often used in pharmaceutical industries. We find it in several medication or molecules in progress.

First, NAPP is encountered in molecules which are ligands of cannabinoid receptors. These latters present an interest in the obesity treatment and in the psychiatric and neurological disorders treatments. These molecules are actually studied by many chemical groups such as Sanofi-Aventis [1], Solvay

[2-4], Merck [5,6], Bayer [7, 8], Pfizer [9], Astrazeneca [10,11], Bristol-Myers Squibb [12].

Second, we meet this hydrazine in α-substituted benzylnaphtyl and benzylthiophene derivates, used for the treatment of diseases such as osteoporosis, breast cancer, and uterine and endometrial fibrosis. For instance, NAPP is employed as potential intermediate in a Raloxifene[o]analogue [13].

NAPP also appears in the synthesis of pyridine-2-carboxamide derivatives which are applied in cancer treatment [14].

In view of the great potential of this hydrazine, we have decided to propose a new synthesis strategy by employing the hydroxylamine-O-sulfonic acid (HOSA) reagent. We also defined the extraction steps in order to propose a relevant purification process. In conformity with the particularly restricting pharmaceutical applications, this latter requires then the study of the solid-liquid-liquid and liquid-vapour binary and ternary phase diagrams, in order to optimize the successive unitary extraction operations.

RESULTS AND DISCUSSION

NAPP ($C_5H_{10}NNH_2$) is obtained by reaction between piperidine ($C_5H_{10}NH$ or PP) and HOSA (NH_2OSO_3H):

Kinetic studies of formation and degradation were carried out in basic medium in order to neutralize the released sulphuric acid [15].

The constant of formation k_1 and oxidation k_2 of NAPP in basic medium at 293 K are as follows:

$k_1 = 3.39 \pm 0.10$ M^{-1}·min^{-1}

$k_2 = 1.17 \pm 0.04$ M^{-1}·min^{-1}

Formation and oxidation reactions were found to be of first order.

The influence of the temperature was studied from 283 to 303 K, showing that the variation of k_1 versus temperature obeys to the Arrhénius law.

We have also proposed a kinetic model, taking into account all the reactions of formation and degradation of the mixture:

$$NH_2OSO_3H + OH^- \rightarrow NH_2OSO_3^- + H_2O \tag{1}$$

$$NH_2OSO_3^- + C_5H_{10}NH \xrightarrow{k_1} C_5H_{10}NNH_2 + HSO_4^- \tag{2}$$

$$OH^- + HSO_4^- \rightarrow SO_4^{2-} + H_2O \tag{3}$$

$$C_5H_{10}NNH_2 + NH_2OSO_3^- \xrightarrow{k_2}$$
$$C_5H_{10}N^+ = N^- + NH_4^+ + SO_4^{2-} \tag{4}$$

$$OH^- + NH_4^+ \xrightarrow{k_2'} NH_3 + H_2O \tag{5}$$

$$2C_5H_{10}N^+ = N^- \rightarrow C_5H_{10}NN = NNC_5H_{10} \tag{6}$$

The resolution of this model shows that the yield of NAPP only depends on two variables. The first one is the p ratio of the initial molar concentrations $[PP]_0/[HOSA]_0$ and the second one, the ratio of rate constants k_2/k_1.

This reaction model was validated in concentrated medium ($1.2 < p < 12$; $283 < T < 373$ K). **Figure 1** shows a modelled kinetic behavior of a mixture with industrial concentrations.

The optimal conditions for the synthesis were then determined as follows:

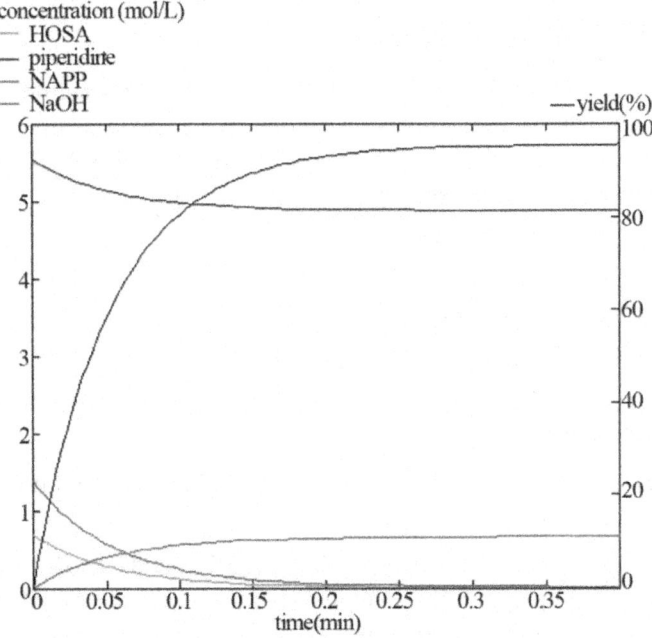

Figure 1. Evolution of the modelled instantaneous concentrations during the synthesis, in basic medium and in industrial conditions ($[HOSA]_0 = 0.69$ M; $[PP]_0 = 5.55$ M; $[NaOH]_0 = 1.39$ M; $T = 293$ K; $p = 8$).

- Aqueous solution of HOSA 32 w% (0.69 M)
- Aqueous solution of PP 66 w% (5.55 M)
- Sodium hydroxide (1.39 M)
- Synthesis temperature: 293 K

The reaction is almost ended after 0.35 second and gives NAPP with 96% yield.

A kinetic study of the formation of NAPP was also realized in buffer medium. The synthesis of NAPP in buffer medium implies the two following parallel reactions:

$$NH_2OSO_3H + C_5H_{10}NH \xrightarrow{k_3} NH_2OSO_3H + C_5H_{10}NH$$

$$2C_5H_{10}NH + H_2SO_4 \rightarrow (C_5H_{10}NH_2)_2 SO_4$$

Leading to the global reaction:

$$NH_2OSO_3H + 3C_5H_{10}NH \rightarrow$$
$$C_5H_{10}NNH_2 + (C_5H_{10}NH_2)_2 SO_4$$

The pH is thus controlled by the buffer mixture piperidine/piperidinium sulphate salt. During the synthesis, 3 moles of PP are consumed to give one mole of NAPP and one mole of piperidinium sulphate salt. It involves the theoretical stoiechiometric relations:

$$-\frac{d[HOSA]}{dt} = -\frac{1}{3}\frac{d[PP]}{dt} = \frac{d[NAPP]}{dt}$$

The rate constant k_3 of formation of NAPP in buffer medium at 293 K was then determined: $k_3 = 2.56 \pm 0.08\,M^{-1}\cdot min^{-1}$ at 293 K.

The formation of a piperidinium salt leads to a slow down in the synthesis reaction of NAPP.

Figure 2 shows the evolution of NAPP yield in basic and buffer media as function of the p ratio.

This graph shows that, when p ≥ 8, the NAPP yields are the same. So, it is also possible to implement this synthesis in buffered medium with the optimal conditions obtained in basic medium.

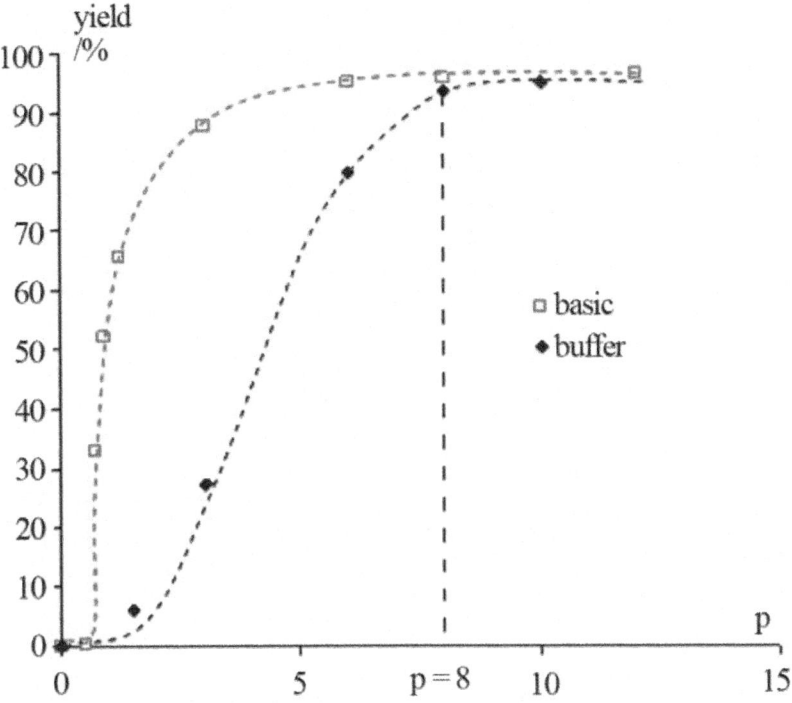

Figure 2. Evolution of the NAPP yield in basic and buffer media as function of the p ratio (with $p = [PP]_0/[HOSA]_0$).

We decided to synthesize NAPP in buffer medium with the following conditions:

- Aqueous solution of HOSA 32 w% (0.69 M)
- Aqueous solution of PP 66 w% (5.55 M)
- Synthesis temperature: 293 K

At the end of the synthesis, the crude solutions are complex and diluted. The extraction process was optimized thanks to the study of the phase diagrams involved. The first step of the extraction consists of a demixing generated in-situ by the sodium sulphate released after neutralization of the piperidinium salt by sodium hydroxyde. Four isothermal sections [16] of the solid-liquidliquid H_2O-Na_2SO_4-$C_5H_{10}NH$ ternary diagram were established at 293, 298, 313 and 323 K under atmospheric pressure by Isoplethic Thermal Analysis (ITA) [17]. A polythermic diagram was then designed between 293 and 323 K [18]. Figures 3(a) and 3(b) show respectively the isotherm 293 and 313 K expressed in mass fractions, under atmospheric pressure.

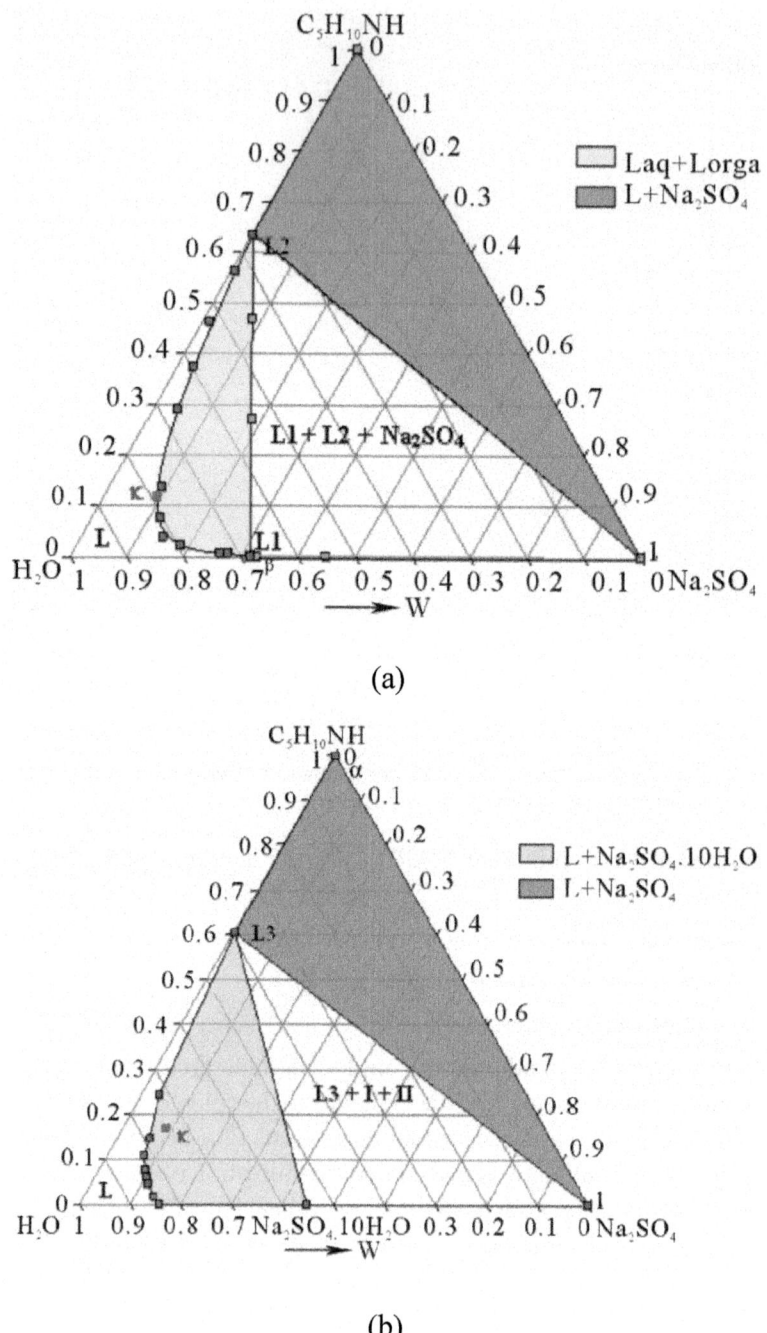

(a)

(b)

Figure 3. (a) Ternary system H_2O-Na_2SO_4-$C_5H_{10}NH$: isotherm 293 K (mass fractions), P = 1 bar. I: $Na_2SO_4 \cdot 10H_2O$; II: Na_2SO_4. (b) Ternary system H_2O-Na_2SO_4-$C_5H_{10}NH$:

isotherm 313 K (mass fractions), P = 1 bar.

The second step is a demixing induced by addition of sodium hydroxide. This one allows concentrating the organic compounds in order to optimize the next distillation operations. This step required thus the study of the solid-liquid-liquid H_2O-NaOH-C_5H_{10}NH ternary diagram. Three isothermal sections were established at 293, 313 and 323 K by ITA [19]. Figures 4(a) and 4(b) show respectively the isotherm 293 and 323 K expressed in mass fractions, under atmospheric pressure.

The final step of the extraction process consists of a distillation of the recovered organic phase mixture composed of water, PP and NAPP (46.5 w% water; 45.6 w% PP; 7.9 w% NAPP). In order to optimize this step, we studied the liquid-vapour H_2O-C_5H_{10}NH-C_5H_{10}NNH$_2$ ternary diagram. The three limit binary systems, $H_2OC_5H_{10}$NH, H_2O-C_5H_{10}NNH$_2$ and C_5H_{10}NH-C_5H_{10}NNH$_2$, were established by ebulliometry under atmospheric pressure. The liquid-vapour H_2O-C_5H_{10}NH-C_5H_{10}NNH$_2$ ternary diagram is characterized by two distillation domains, linked to the existence of a positive azeotropic mixture in the H_2O-C_5H_{10}NH binary system. **Figure 5** shows the H_2O-C_5H_{10}NH-C_5H_{10}NNH$_2$ polythermal ternary system under atmospheric pressure.

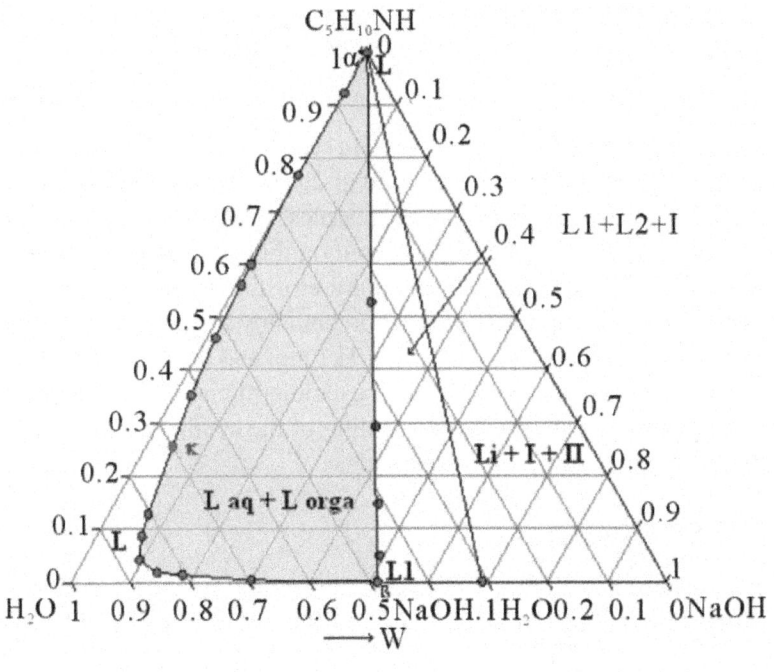

(a)

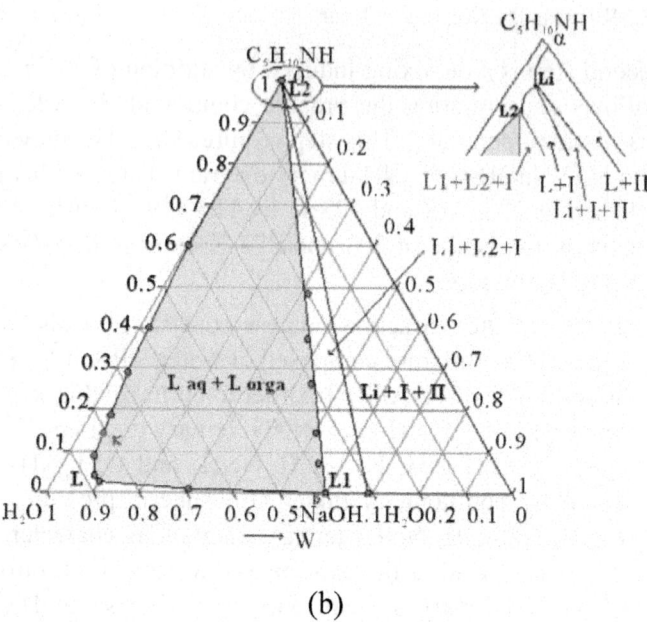

(b)

Figure 4. (a) Ternary system H_2O-NaOH-C_5H_{10}NH: isotherm 293 K (mass fractions), $P = 1$ bar, I: NaOH·H_2O, II: NaOH. (b) Ternary system H_2O-NaOH-C_5H_{10}NH: isotherm 293 K (mass fractions), $P = 1$ bar, I: NaOH·H_2O, II: NaOH.

The experimental coordinates of the H_2O-PP azeotropic mixture are:

$x(C_5H_{10}NH) = 0.661$ $T_{vap} = 364.95$ K To secure the process, two measurement sets of mixing enthalpies were realized at 298 K, by mixing calorimetry. The first series concerns the solvation enthalpies of HOSA-water and PP-water binary mixtures. The second one, focused on the measurement of the reaction enthalpy of NAPP according to the dilution of the reagent involved, allowed us to determine the hypothetical reaction enthalpy in anhydrous medium. The infinite dilution enthalpy is equal to -140 kJ/mol and the enthalpy in anhydrous medium is estimated at -340 kJ/mol. The exothermicity of the reaction requires thus the necessity of an impressive cooling of the synthesis reactor in order to manage the temperature of the medium.

PROCESS DESIGN

We can propose two different processes depending on whether PP is injected anhydrous or as an azeotropic mixture of PP in water (66 w% PP).

In both cases, the first step consists in carrying out the synthesis reaction at 293 K, in a buffer medium and with a molar ratio of initial concentrations $[C_5H_{10}NH]_0/[NH_2OSO_3H]_0$ equals to 8. The NAPP yield reaches then 93%.

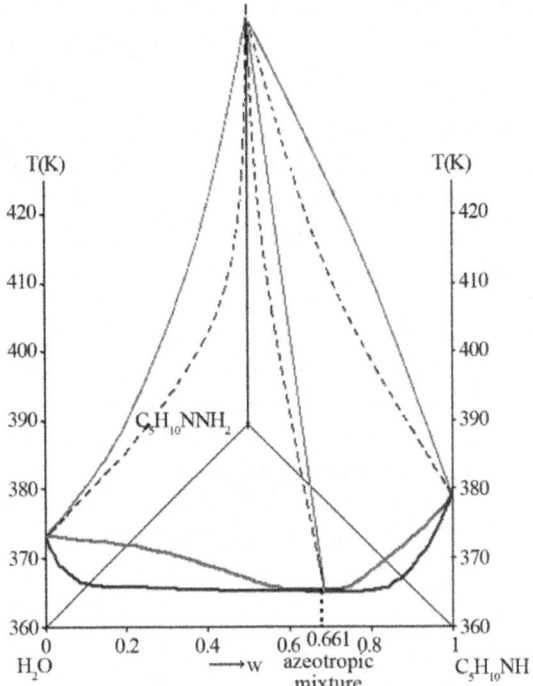

Figure 5. Liquid-vapour H_2O-$C_5H_{10}NH$-$C_5H_{10}NNH_2$ ternary system, P = 1 bar (mass fractions).

After the synthesis, the crude mixture is mainly composed of NAPP, piperidinium sulphate salt, PP in excess, water and some by-products (tetrazene and triazanium salt derivatives). In a second operation, we add a 32 w% aqueous solution of sodium hydroxide which generates sodium sulphate by neutralization of the piperidinium salt and induced the salting-out.

G_{anh} is the mass composition of the crude mixture obtained after the NAPP synthesis by injection of anhydrous PP and neutralization of the salt. G_{az} is the mass composition of the crude mixture after addition of an azeotropic mixture of PP and water and neutralization of the salt at the end of the NAPP synthesis. These two compositions are the following:

G_{anh} = 44.7 w% PP; 12.5 w% Na_2SO_4; 42.8 w% H_2O.

G_{az} = 33.1 w% PP; 9.9 w% Na_2SO_4; 57 w% H_2O.

As shows the isotherm 293 K of the solid-liquid H_2ONa$_2SO_4$-$C_5H_{10}NH$ ternary diagram (**Figure 6**), the G_{az} and G_{anh} points are located respectively in the L + $Na_2SO_4 \cdot 10H_2O$ and the L3+ $Na_2SO_4 \cdot 10H_2O$ + Na_2SO_4 domains.

At this temperature, any demixing zone appears.

But at 313 K (**Figure 7**), G_{az} and G_{anh} are respectively located in the demixing zone and in the $L1 + L2 + Na_2SO_4$ triphasic domains.

It is thus preferable to synthesize NAPP starting from an azeotropic mixture of PP and water and implement the neutralization of the piperidinium salt at 313 K. In these conditions there is no precipitation of anhydrous sodium sulphate. After the neutralization step, the recovered organic liquid M is mainly composed of water (46.5 w%), PP (45.6 w%) and NAPP (7.9 w%).

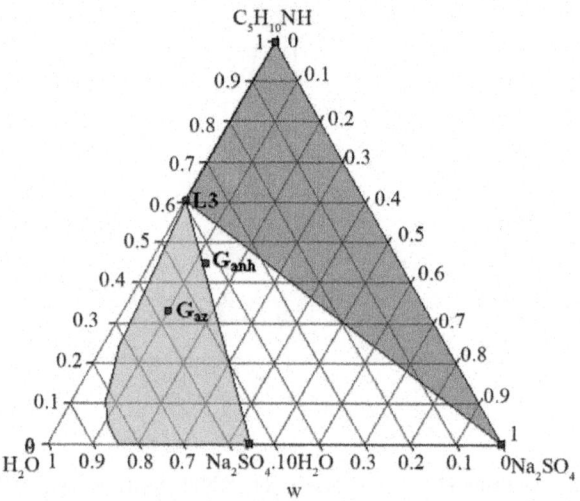

Figure 6. Isotherm 293 K of the solid-liquid H_2O-Na_2SO_4- $C_5H_{10}NH$ ternary diagram, P = 1 bar. Location of the G_{anh} and G_{az} crude mixtures.

Looking at the liquid-vapour ternary system, the mixture M is in fact located in the water/azeotrope/NAPP distillation field (**Figure 8**).

In this configuration, two successive distillation operations are required to get rid of the azeotropic mixture and the water in excess and finally a third distillation, under reduced pressure, permits to obtain the product in conformity with the pharmaceutical applications (purity up to 99 w%).

In order to simplify the distillation process, we preferred to dehydrate partially the organic liquid M thanks to a demixing observed in the solid-liquid-liquid H_2ONaOH-$C_5H_{10}NH$ ternary system.

The aim of this step is to generate a new demixing by addition of an aqueous solution of sodium hydroxide (50 %w). The added quantity of soda was optimized in order to obtain an organic phase with a mass composition very close to those of the azeotrope. This organic liquid M' is then composed of water (31.8 w%), PP (59.2 w%) and NAPP (9 w%).

This mixture M' is very near the monoazeotropic distillation curve (**Figure 9**).

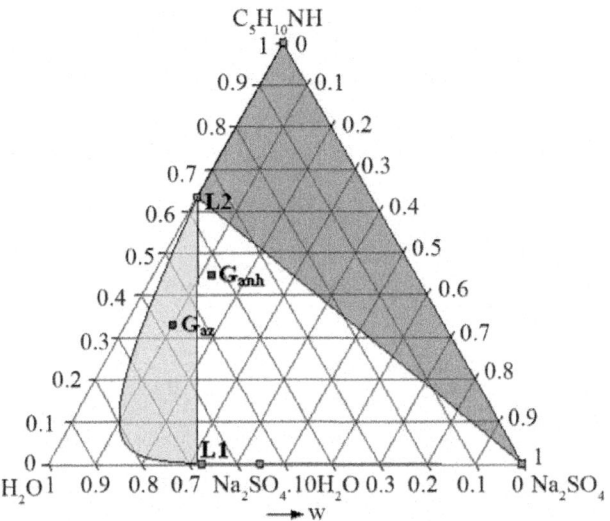

Figure 7. Isotherm 313 K of the solid-liquid H_2O-Na_2SO_4- $C_5H_{10}NH$ ternary system, P = 1 bar. Location of the G_{anh} and G_{az} crude mixtures.

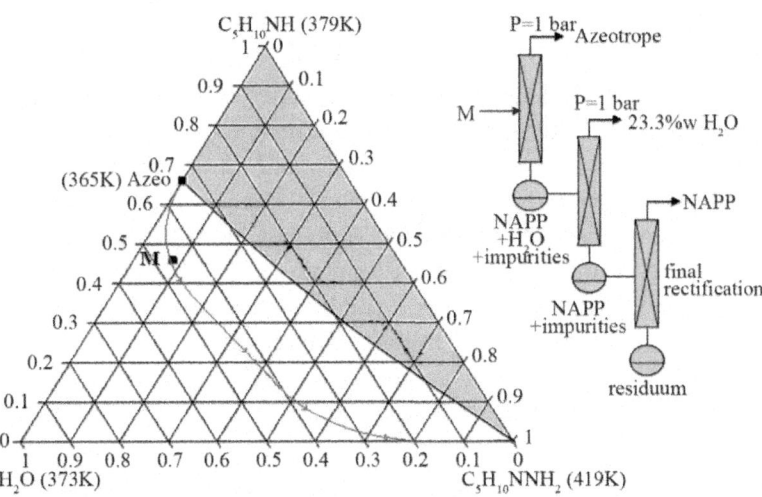

Figure 8. Distillation trajectory of the organic liquid M (water 46.5 w%, PP 45.6 w%, NAPP 7.9 w%).

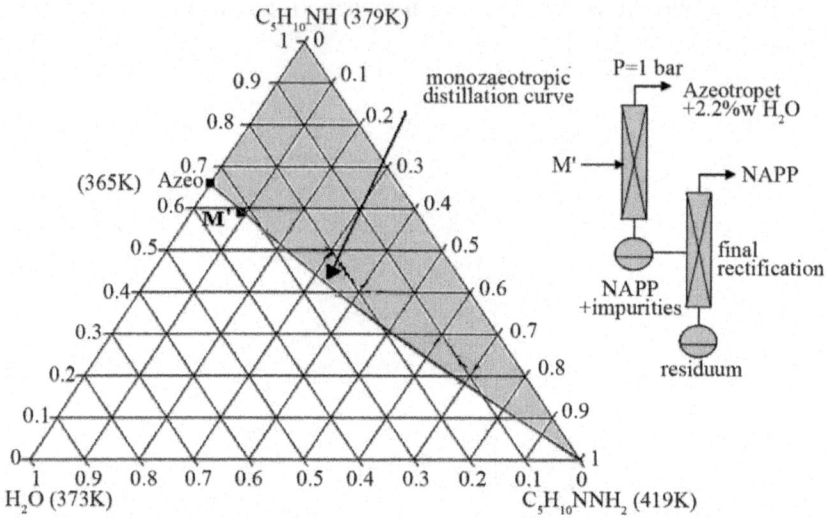

Figure 9. Distillation trajectory of the organic liquid M' (water 31.8 w%, PP 59.2 w%, NAPP 9 w%).

In these optimized conditions, two successive distillations are thus necessary to obtain anhydrous NAPP. The small excess of water has no impact on the final yield of NAPP. The azeotropic mixture with 2.2 w% of water can be directly injected upstream from the synthesis (**Figure 10**).

For the anhydrous version process (**Figure 11**), we can totally dehydrate the organic liquid M thanks to a demixing observed in the solid-liquid-liquid H_2O-NaOHC$_5$H$_{10}$NH ternary system. Experimentally, this step is fastidious and requires several runs of soda additions and filtrations. At the end of this step, we obtain then an organic phase mainly composed of PP and NAPP and it becomes easy to separate the two compounds by distillation. However, the organic phase is not anhydrous and contains a small amount of water. So, during the distillation, we recover a small fraction of the H_2O-PP azeotropic mixture, which is economically necessary to retreat.

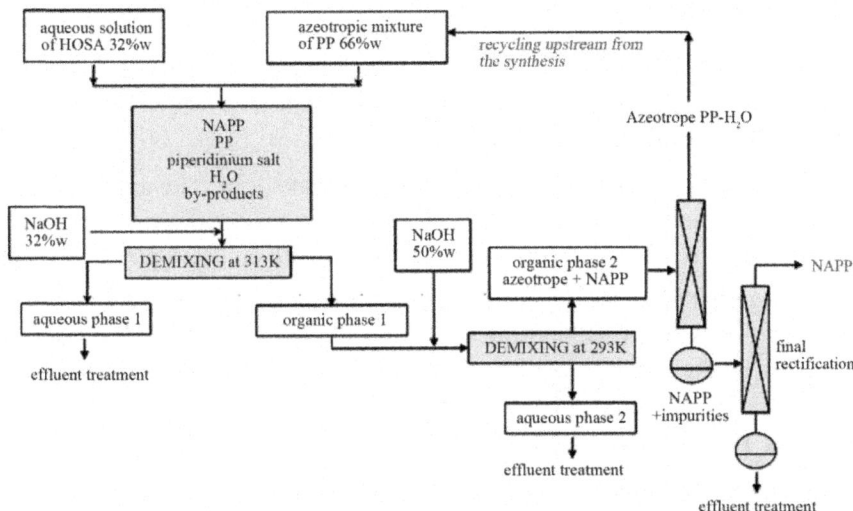

Figure 10. Process with PP-water azeotropic mixture.

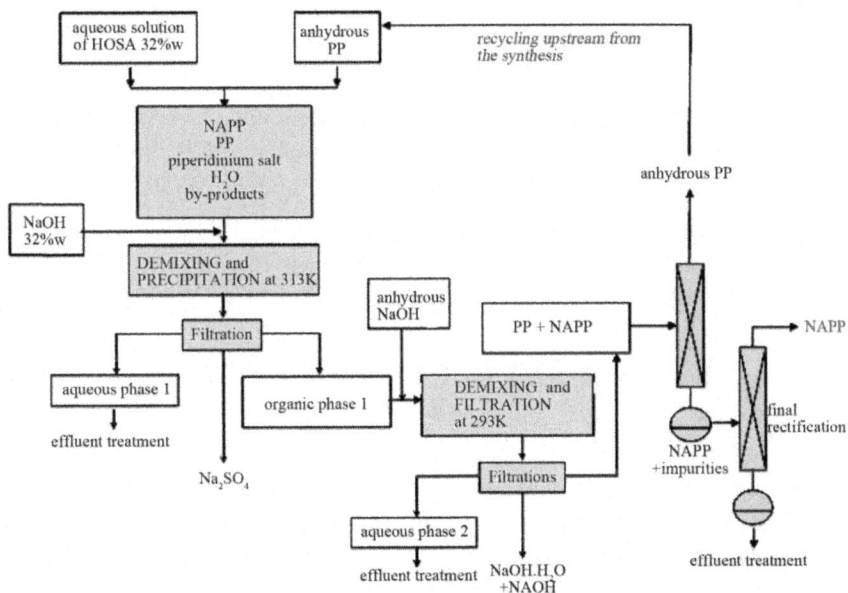

Figure 11. Process with anhydrous PP.

The synthesis using azeotropic mixture of water and PP is more adapted, for a batch or a continuous process, with a minimum of extraction operations and all steps are carried out in fluid phase.

CONCLUSIONS

The synthesis of NAPP was optimized thanks to kinetic studies. In order to obtain a 96% yield of NAPP, the synthesis is realized at 293 K with an excess of PP (p = 8), in buffer medium.

The extraction parameters were defined thanks to a relevant exploitation of the phase diagrams involved. It consists of two successive demixings and two distillations. The final product is anhydrous and contains any by-product.

Two global processes of synthesis and extraction of NAPP were developed, one using anhydrous PP and the other, the PP-H_2O azeotropic mixture. The process with the azeotropic mixture seems to be the most relevant and flexible.

ACKNOWLEDGEMENTS

Thanks to the Région Rhone-Alpes FRANCE: Study funded by EMERGENCE Project 2005-2008.

REFERENCES

1. F. Barth and P. Casellas, "Substituted N-Piperidino 3-Pyrazolecarboxamide," European Patent 0656354, 1995.

2. C. G. Kruse, J. H. M. Lange, A. H. J. Herremans and H. H. Van Stuivenberg, "1H-Imidazole Derivatives Having CB_1 Agonistic, CB_1 Partial Agonistic or CB_1-Antagonistic Activity," International Patent WO03027076, 2003.

3. C. G. Kruse, J. H. M. Lange, T. Jacobus, A. H. J. Herremans and H. H. Van Stuivenberg, "Novel 4,5-Dihydro- 1H-pyrazole Derivatives Having CB_1-Antagonistic Activity," International Patent WO03026647, 2003.

4. A. R. Stoit, J. H. M. Lange, A. P. Den Hartog, E. Ronken, K. Tipker, H. H. Van Stuivenberg, J. A. R. Dijksman, H. C. Wals and C. G. Kruse, "Design, Synthesis and Biological Activity of Rigid Cannabinoid CB_1 Receptor Antagonists," Chemical and Pharmaceutical Bulletin, Vol. 50, No. 8, 2002, pp. 1109-1113. doi:10.1248/cpb.50.1109

5. P. E. Finke, S. G. Mills, C. W. Plummer, S. K. Shah and Q. T. Truong, "Substituted Imidazoles as Cannabinoid Receptor Modulators," International Patent WO03007887, 2003.

6. W. K. Hagmann, H. Qi and S. K. Shah, "Substituted Imidazoles as Cannabinoid Receptor Modulators," International Patent WO03063781, 2003.

7. R. A. Smith, S. J. O'Connor, S. N. Wirtz, C. Wong Wai, S. Choi, H. C. Kluender, N. Su, G. Wang, F. Achebe and S. Ying, "Imidazole-4-

carboxamide Derivatives, Preparation and Use Thereof for Treatment of Obesity," International Patent WO03040107, 2003.

8. R. A. Smith, H. C. Kluender, N. Su, R. C. Lavoie and J. Fan, "Preparation and Use of Pyrrole Derivatives for Treating Obesity," International Patent WO03027069, 2003.

9. A. L. Hildebrandt, D. M. Kelly-Sullivan and S. C. Black, "Antiobesity Effects of Chronic Cannabinoid CB_1 Receptor Antagonist Treatment in Diet-Induced Obese Mice," European Journal of Pharmacology, Vol. 462, No. 1-3, 2003, pp. 125-132. doi:10.1016/S0014-2999(03)01343-8

10. J. M. Wilsterman and A. I. K. Berggren, "Pyrazine Compounds and Pharmaceutical Compositions Containing them," International Patent WO03051850, 2003.

11. A. I. K. Berggren, S. J. Bostrom, S. T. Elebring, P. Greasley, E. Terricabra and J. M. Wilstermann, "5,6-Diaryl-pyrazine-2-amide Derivatives as CB_1 Antagonists," International Patent WO03051851, 2003.

12. S. Liu, "N-Substituted 3-Hydroxy-4-pyridinones and Pharmaceuticals Containing Thereof," International Patent WO03065991, 2003.

13. B. S. Muehl, "3-Benzyl-benzothiophenes," US Patent 6417199, 2002.

14. J. Dumas, W. Lee, Y. Chen, L. Adnane, W. Scott, S. Verna, J. Chen, Z. Chen and L. Yi, "Substituted Pyridine Derivatives Useful in the Treatment of Cancer and Other Disorders," European Patent 1603879, 2005.

15. E. Labarthe, A. J. Bougrine, V. Pasquet and H. Delalu, "Kinetic Modelling of Synthesis of N-Aminopiperidine from Hydroxylamine-O-sulfonic Acid and Piperidine," Kinetics and Catalysis, Vol. 53, No. 1, 2012, pp. 25-35. doi:10.1134/S0023158412010041

16. E. Labarthe, A. J. Bougrine, H. Delalu, J. Berthet and J. J. Counioux, "Equilibrium Study between Condensed Phases by Isoplethic Thermal Analysis When a Miscibility Gap Is Observed," Journal of Thermal Analysis and Calorimetry, Vol. 95, No. 1, 2009, pp. 135-139. doi:10.1007/s10973-008-9000-8

17. J. Berthet and J. J. Counioux, "Procédé et Dispositif de Mesure de la Solubilité d›au Moins un Soluté Solide, Liquide ou Gazeux dans un Solvant, Par Analyse Thermique Isopléthique," French Patent 9313402, 1993.

18. E. Labarthe, A. J. Bougrine, J. Berthet and H. Delalu, "Study of the Polythermal Diagram Water-Sodium Sulphate-Piperidine," Journal of Themal Analysis and Calorimetry, Vol. 100, No. 3, 2010, pp. 1099-1105. doi:10.1007/s10973-009-0567-5

19. E. Labarthe, A. J. Bougrine and H. Delalu, "Extraction Optimization of Organic Compounds by Demixing Observed in the H₂O-NaOH-Piperidine Ternary Diagram," Journal of Themal Analysis and Calorimetry, Vol. 102, No. 3, 2010, pp. 1119-1122. doi:10.1007/s10973-010-0744-6

Chapter 5

IMMOBILIZATION OF COMMERCIAL INULINASE ON ALGINATE–CHITOSAN BEADS

Juliano Missau, Amir J Scheid, Edson L Foletto, Sergio L Jahn, Marcio A Mazutti and Raquel C Kuhn
Department of Chemical Engineering, Federal University of Santa Maria, Santa Maria 97105-900, Brazil

ABSTRACT

The commercial inulinase obtained from *Aspergillus niger* was effectively immobilized on alginate-chitosan beads which were hardened with glutaraldehyde. The immobilization conditions were studied using Plackett & Burmann experimental design and central composite rotational design (CCRD). The effects of chitosan, glutaraldehyde, sodium alginate and calcium chloride concentrations in order to obtain a better immobilization yield were optimized. In the Plackett & Burman experimental design, the sodium alginate and calcium chloride had a significant effect ($p < 0.1$), but only the calcium chloride showed a positive effect, indicating that as higher the concentration, better is the immobilization yield. In the central composite rotational design (CCRD), the best results were obtained in the central point, using sodium alginate (1% w/v) and calcium chloride (4% w/v) as conditions for inulinase immobilization. By the CCRD, the optimal immobilization strategy was: chitosan (0.1% w/v), glutaraldehyde (0.1% v/v), sodium alginate (1% w/v) and calcium chloride (4% w/v). In this condition, the enzyme loading capacity was 668 U/g gel beads and the effect of temperature on the immobilized enzyme activity was also evaluated, showing better activity at 50°C. The immobilized enzyme maintained 76% of its activity in six days at room temperature.

INTRODUCTION

Inulinases are enzymes potentially useful on the production of high fructose syrups (HFS) by enzymatic hydrolysis of inulin, conducting to a yield of 95% [1]. Inulinases are enzymes widely used for the production of fructooligosaccharides, compounds with functional and nutritional properties for use in low-calorie diets, stimulation of *Bifidus* and as a source of dietary fiber in food preparations [2].

Enzyme immobilization increases the catalytic properties of enzyme, allow the continuous reuses of costly enzyme to make it economically viable for industrial applications [3, 4]. The enzymes immobilization is usually carried out by three methods: covalent binding to a supports, adsorption of enzyme molecules on a support material and entrapment or encapsulation of enzyme in polymers. The covalent binding and adsorption methods both have disadvantages because they have possibility to affect the substrate binding site of enzyme and enforce diffusion limitation on the enzyme which ultimately causes the decrease in enzyme activity [4]. Entrapment is one step process in which chances of activity lost is comparatively low. Polymers such as alginate were used for entrapment of enzymes [4, 5].

Calcium alginate hydrogel beads are commonly used carriers in the entrapment immobilization of biocatalyst [5] owing to their significant advantages such low cost, high porosity, and simplicity of preparation, however, this material has some limitations these are due to biocompatibility, including high biomolecule leakage, and large pore size [5, 6]. For the encapsulation efficiency and control release of enzyme from the gel matrix, the covalent cross-linking with polymers, such as chitosan, and coating the surface of alginate gel beads with other reagents, such as glutaraldehyde, have been used [5].

This study was based on immobilization of commercial inulinase from *Aspergillus niger* within alginate–chitosan beads. In order to obtain a better immobilization yield, the immobilization parameters such as chitosan, glutaraldehyde, sodium alginate and calcium chloride concentrations were optimized. The Plackett & Burman and central composite rotational design (CCRD) were employed to evaluate the effects of immobilization parameters. Thermal and storage stabilities were also evaluated in this work.

MATERIAL AND METHODS

Material

The commercial inulinase was purchased from Sigma–Aldrich, which was obtained from *Aspergillus niger* (Fructozyme, exo-inulinase EC 3.2.1.80 and endo-inulinase EC 3.2.1.7). The chitosan was purchased from Purifarma (Brazil) and others reagents from Vetec (Brazil).

Enzyme Immobilization

For the inulinase immobilization protocol, a methodology adapted from Zhou et al. [5] was used: alginate was dissolved in water and the equal volume inulinase enzyme solution (1:100, enzyme:acetate buffer) was added by mild shaking on a rotary shaker. Chitosan was ultrasonically dispersed in an acetic acid solution (5% v/v) for 1 h and $CaCl_2$ solution was added. Alginate/inulinase mixture was extruded dropwise through a peristaltic pump into a 50 mL chitosan/$CaCl_2$ solution and hardened in this solution. The formed spherical beads were rinsed with sterile NaCl solution (0.9% w/v) (2×20 mL) and then treated with glutaraldehyde solution for 2 h. The immobilized inulinase was washed thrice with sterile distilled water and then directly used for the measurements of activity and stability.

Inulinase Activity Assay

An aliquot of 0.5 g of the enzyme was incubated with 4.5 mL sucrose solution (2% w/v) in sodium acetate buffer (0.1 M, pH 4.8) at 50°C. Reducing sugars released were measured by the 3.5-dinitrosalicylic acid method [7]. A separate blank was set up for each sample to correct the non-enzymatic release of sugars. One unit of inulinase activity was defined as the amount of enzyme necessary to hydrolyze 1 μmol of sucrose per minute under the mentioned conditions (sucrose as a substrate). Results were expressed in terms of inulinase activity (enzyme loading) per gram of gel beads (U/g).

At each step, the pellets were washed with distilled water to remove the excess of glutaraldehyde and unbound enzyme. The immobilization yield (Y%) was determined as defined in the Equation 1:

$$Y\ (\%) = \frac{activity\ immobilization\ enzyme}{activity\ free\ enzyme} \times 100 \quad (1)$$

The amount of protein loaded on the support was calculated from the difference of initially added protein to the protein obtained in the washing plus supernatant. The protein content was determined by Biuret method using bovine serum albumin as a standard [8].

The immobilization efficiency was defined in the Equation 2:

$$E\ (\%) = \frac{specific\ activity\ bound\ to\ the\ support}{specific\ activity\ initially\ added} \times 100 \quad (2)$$

Experimental Design

The effects of chitosan (0.1-0.4% w/v), glutaraldehyde (0.1-0.9% v/v), sodium alginate (1-5% w/v) and calcium chloride (1-3% w/v) were assessed by means a Plackett & Burman for four independent variables. Table 1 presents the range of investigated variables. The significance variables in the Plackett & Burman were studied in the central composite rotational design (CCRD). All the results were analyzed using the software Statistica® 8.0 (Statsoft Inc., Tulsa, USA).

Table 1: Matrix of Plackett & Burman experimental design

Independent variables	Coded levels of variables		
	-1	**0**	**+1**
Chitosan (%)	0.1	0.25	0.4
Glutaraldehyde (%)	0.1	0.50	0.9
Sodium alginate (%)	1	3	5
$CaCl_2$ (%)	1	2	3

Thermal Stability

The stability was determined by incubation of immobilized enzyme in 0.1 M acetate buffer (pH 4.8) without substrate at 30, 50 and 70°C. Samples were taken at different intervals during 4 hours and the inulinase activity was determined. The relative activity at each temperature was determined by taking the activity at 0 min as 100%.

Shelf Stability

The shelf stability was determined by incubation of immobilized enzyme in 0.1 M acetate buffer (pH 4.8) without substrate at room temperature (25°C). Samples were taken at different intervals and the inulinase activity was determined. The relative activity at each temperature was determined by taking the activity at 0 min as 100%.

RESULTS AND DISCUSSION

Plackett & Burman

In large-scale processes, the enzyme can be immobilized and the process cost is very important. Therefore, the conditions for the inulinase immobilization were studied in this work. The enzyme immobilization on alginate beads is not only inexpensive, but also used in mild conditions [5]. So, the sodium alginate has been considered for the entrapment of enzymes due many advantages [5].

In the preliminary experiments, the main variables (chitosan, glutaraldehyde, calcium chloride, sodium alginate) that could influence on the immobilization yield were studied through Plackett & Burman experimental design for the optimization of inulinase immobilization. Table 2 and Pareto Chart (Figure 1) showed that only the sodium alginate and calcium chloride had a significant effect at 90% of confidence ($p < 0.1$) on the inulinase immobilization yield. The sodium alginate had a negative effect, meaning that as lower the sodium alginate concentration, better is the immobilization yield. On the other hand, the calcium chloride showed a positive effect, indicating that as higher the concentration, better is the immobilization yield. The chitosan and glutaraldehyde were not statistically significant within the levels studied.

Table 2: Effects of Plackett & Burman experimental design for immobilization yield

	Effects	**Error**	**t(6)**	**p-value**
Mean	27.125	0.896	30.261	<0.0000008
Chitosan (%)	3.858	2.102	1.835	0.1161
Glutaraldehyde (%)	1.851	2.102	0.881	0.4124
Sodium alginate (%)	-7.531	2.102	-3.582	0.01161
Calcium chloride (%)	6.637	2.102	3.157	0.01963

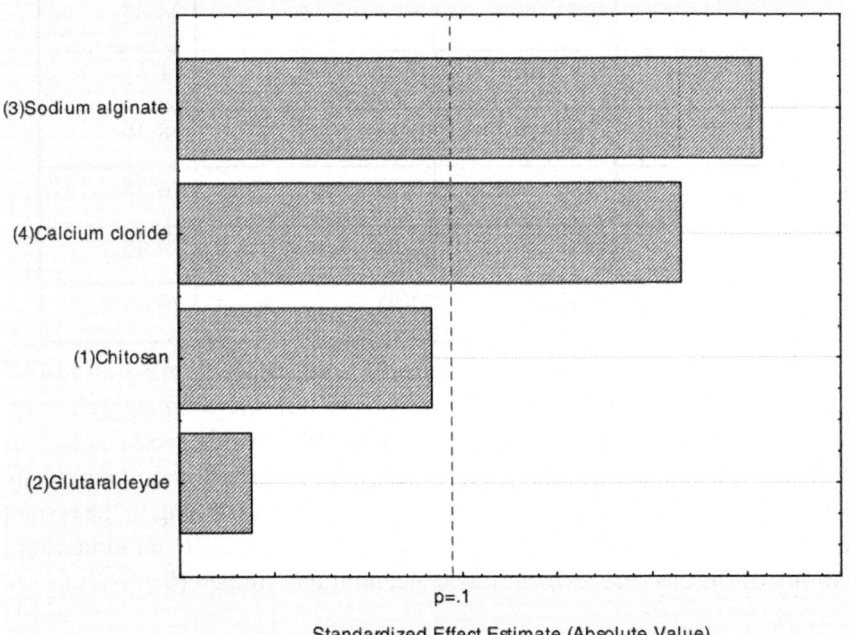

Figure 1: Pareto Chart for the variables independents in the Plackett & Burman experimental design.

Considering the Plackett & Burman experimental design results, an additional set of experiments were carried out according to a central composite rotational design (CCRD) in order to find the optimal conditions for the inulinase immobilization. The variables real and coded were presented in the Table 3. The variables chitosan and glutaraldehyde were fixed at 0.1% in the central composite rotational design.

Table 3: Results of inulinase immobilization using central composite rotational design

Assays	Sodium algi-nate	Calcium chloride	Y(%)
1	-1(0.5)	-1(3)	13.58
2	-1(0.5)	+1(5)	25.52
3	+1(1.5)	-1(3)	22.11
4	+1(1.5)	+1(5)	30.73
5	0(1)	-1.41(2.6)	25.34
6	0(1)	+1.41(5.4)	24.45
7	-1.41(0.3)	0(4)	11.11
8	+1.41(1.7)	0(4)	18.10
9	0(1)	0(4)	36.38
10	0(1)	0(4)	39.48
11	0(1)	0(4)	36.63

The glutaraldehyde cross-links enzyme and gelatin forming an insoluble structure. Glutaraldehyde treatment also stabilizes the alginate gel, helping in the prevention of the leakage of enzymes [9]. In this work, it was used to maintain the stable beads. At the concentration studied, the glutaraldehyde showed no significant influence on the response, but according to the results, it was possible to conclude that the glutaraldehyde is especially important for the stability of the enzyme, even at low concentrations studied (0.1% v/v).

Central Composite Rotational Design (2^2)

The better results concerning the CCRD were obtained in the central point (Table 3). According to these results, the best conditions for the inulinase immobilization yield were observed using sodium alginate (1% w/v) and calcium chloride (4% w/v). The effect of calcium chloride concentration is important to secure stable calcium alginate beads with maximum immobilization yield [4].

The enzyme immobilization presented activity values lower than those obtained for the free enzyme. Cheirsilp et al. [10] and Zhou et al. [5] reported a decrease of enzyme activity in immobilized alginate beads, similar to the observed in this work. The decrease in the immobilized enzyme activity could be explained due to diffusional limits, steric effects, structural changes in the enzyme occurring upon covalent coupling, or lowered accessibility of substrate to the active site of the immobilized enzyme [11].

The enzyme loading capacity i.e., enzyme per gram of gel beads, was 668 U/g in the best condition of the CCRD. Danial et al. [12] obtained 530 U/g and 336 U/g gel using one and two-step method on grafted alginate, respectively. The crude inulinase was assayed for its activity and protein content, the specific activity was calculated according Eq. 2 and was 66%.

The results obtained in the central composite rotational design were used to build the quadratic models expressing the inulinase immobilization yield as functions of the independent variables.

Based on statistical analysis of model parameters, one empirical model was presented below (Eq. 3). Coded model (Eq. 3) was used to generate response surface (Figure 2) for the analysis of the variables effects on the immobilization. It presents the significant terms ($p < 0.05$) concerning of inulinase immobilization yield and it was validated by analysis of variance (ANOVA), which was presented in Table 4. According to the ANOVA analysis of the immobilization, the calculated F was about 2.9 times greater than the tabulated ones and the determination coefficient was 0.76. The high values for the determination coefficient indicate good fitting of experimental data, allowing the use of such models to predict process performance. For the biotechnological process, this determination coefficient is acceptable because of the high variability in the bioprocess.

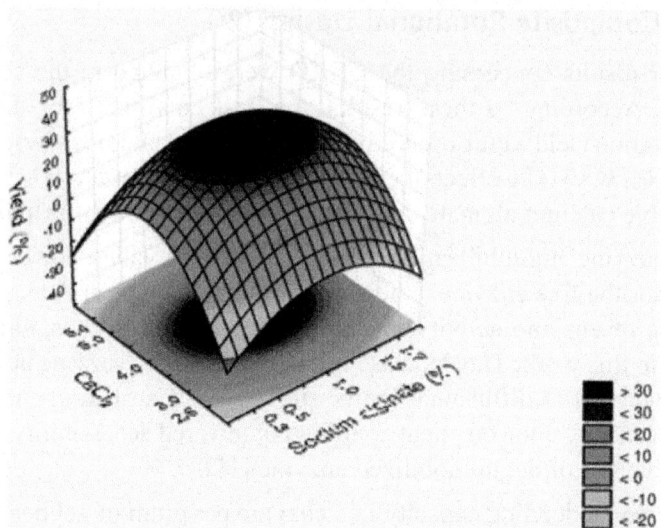

Figure 2: Response surface for the inulinase immobilization according to the central composite rotational design.

Table 4: ANOVA for inulinase immobilization

	SS	*df*	MS	F	R²
Regression	672.26	2	336.13	12.88	76.3
Residual	208.65	8	26.08		
Total	880.91	10			

$F_{0.05;\ 2;\ 8} = 4.46.$

$$Y\ (\%) = 37.48 - 5.49\ X_1{}^2 - 10.66\ X_2{}^2 \qquad (3)$$

Where X_1 is the sodium alginate and X_2 is the calcium chloride.

The validated model was used to optimize the process using the tool response/desirability profiling of Statistica 8.0. The desirability function allows the response surface produced be inspected by fitting the observed responses using the above mentioned equation based on levels of the independent variables. This equation was used to predict values for response (inulinase immobilization yield) at different combinations of levels of the independent variables, specify desirability functions for the dependent variables, and to search for the levels of the independent variables that produce the most desirable responses for the dependent variables (immobilization yield) [13].

Thermal and Shelf Stabilities

The optimum temperature of immobilized enzyme was 50°C. According to Figure 3, it is clear that in the temperature of 50°C the relative activity of immobilized enzyme was considerably higher compared with other temperatures studied (30 and 70°C). Rocha et al. [14] achieved maximum activities for immobilized inulinase onto Amberlite IRC 50 at 50°C. Danial et al. [12] and Yewale et al. [15] found 60°C as the best temperature for the immobilized inulinase on grafted alginate and chitosan, respectively. Richeti et al. [16] observed optimum values for immobilization at 55°C. In this work, in the temperature of 70°C was observed that beads did not maintain the structure and weakened. The results revealed that at 50°C and after 4 hours the immobilized enzyme activity retained 86.5% of its activity. The data shown in Figure 4indicated that the immobilized enzyme retained over 76% of its activity in six days at room temperature. However, this storage stability could be improved at lower temperatures, e.g. 4–5°C. The immobilized enzyme lost practically all of its activity at room temperature after 20 days; the relative activity was 20%.In the Figure 5 was observed that the temperature of 50°C maintained 100% of relative activity during 120 minutes, and at 30°C and 70°C around 80% and 50% of relative activity was maintained, respectively.

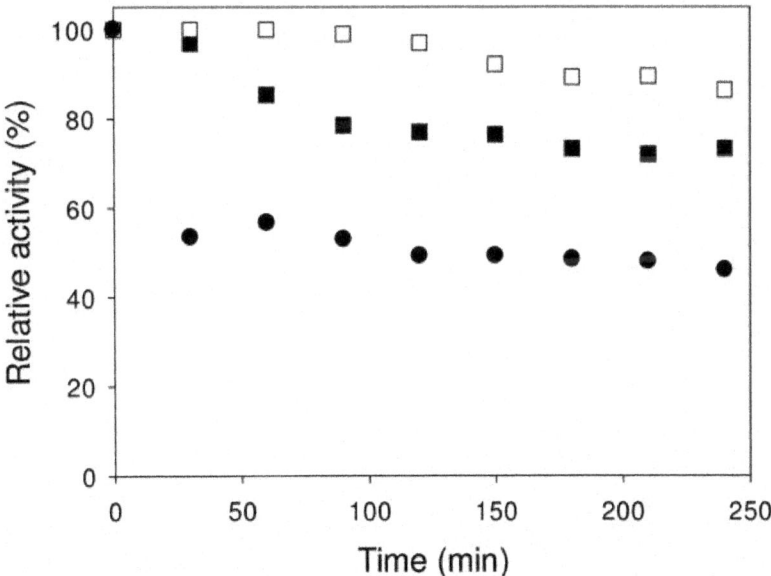

Figure 3: Thermal stability of immobilized enzyme ((■) 30°C, (□) 50°C and (●) 70°C).

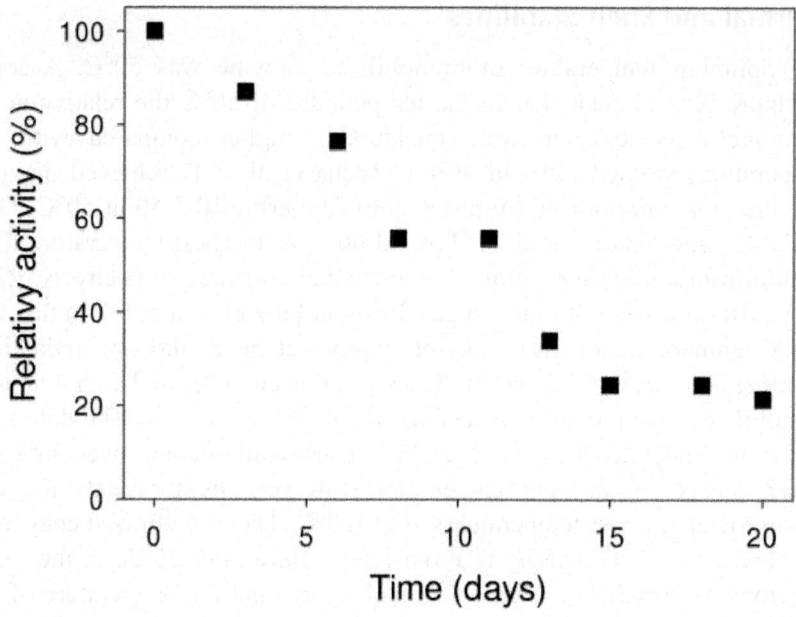

Figure 4: Shelf stability of immobilized inulinase during storage at room temperature.

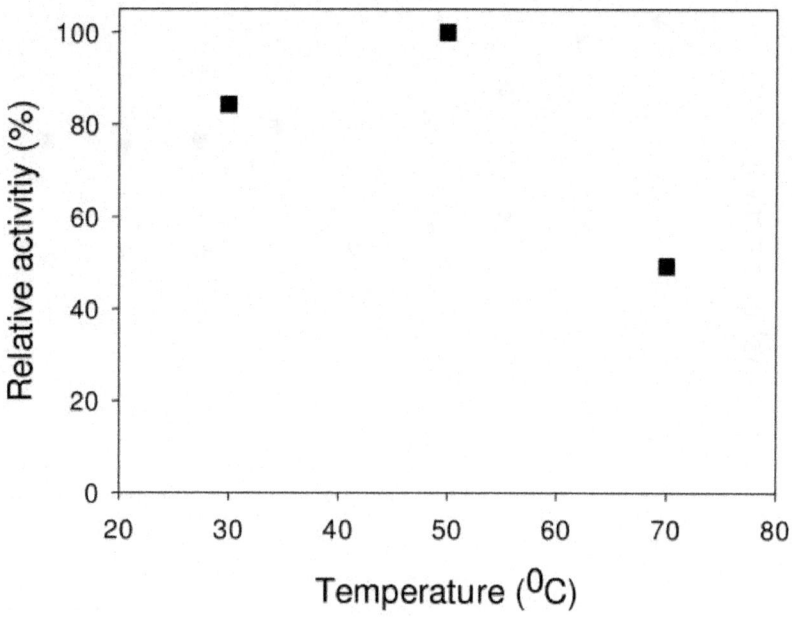

Figure 5: The effect of temperature on enzyme activity.

CONCLUSION

Inulinase immobilization could be carried out successfully using alginate-chitosan beads hardened with glutaraldehyde. By the Plackett & Burman experimental design only the variables sodium alginate and calcium chloride presented significant effect ($p < 0.1$). In the CCRD the optimal immobilization strategy was: chitosan (0.1% w/v), glutaraldehyde (0.1% v/v), sodium alginate (1% w/v) and calcium chloride (4% w/v). In this condition, the optimum temperature in the thermal stability studied was 50°C and the inulinase immobilization retained 86.5% of the relative activity during 240 minutes. The enzyme loading capacity was 668 U/g gel beads, which could be indicating that the inulinase immobilization on the alginate-chitosan beads is a promising technique.

ACKNOWLEDGEMENTS

The authors thank CNPq, CAPES, FAPERGS and Programa FIPE Júnior/UFSM for the financial support of this work and scholarships.

AUTHORS' CONTRIBUTIONS

All authors contributed equally in this work. All authors read and approved the final manuscript.

REFERENCES

1. Ettalibi M, Baratti JC: Sucrose hydrolysis by thermostable immobilized inulinases from *Aspergillus ficcum* . *Enzyme Microb Technol* 2001, 28:596–601.

2. Silva-Santisteban BOY, Maugeri F: Agitation, aeration and shear stress as key factors in inulinase production by *Kluyveromyces marxianus* . *Enzyme Microb Technol* 2005, 36:717–724.

3. Mateo C, Palomo JM, Fernandez-Lorente G, Guisan JM, Fernandez-Lafuente R: Improvement of enzyme activity, stability and selectivity via immobilization techniques. *Enzyme Microb Technol* 2007, 40:1451–1463.

4. Rehman HU, Aman A, Silipo A, Qader SAU, Molinaro A, Ansari A: Degradation of complex carbohydrate: Immobilization of pectinase from *Bacillus Licheniformis* KIBGE-IB 21 using calcium alginate as a support. *Food Chem* 2013, 139:1081–1086.

5. Zhou Z, Li G, Li Y: Immobilization of *Saccharomyces cerevisiae* alcohol dehydrogenase on hybrid alginate-chitosan beads. *Int J Biol Macromol* 2010, 47:21–26.

6. Smidsrod O, Skjak-Brlk G: Alginate as immobilization matrix for cells. *Trends Biotechnol* 1990, 8:71–78.

7. Miller GL: Use of dinitrosalisylic acid reagent for determination of reducing sugar. *Anal Chem* 1959, 31:426–428.

8. Bernardini RD, Harnedy P, Bolton D, Kerry J, O'Neill E, Mullen AM, Hayes M: Antioxidant and antimicrobial peptidic hydrolysates from muscle protein sources and by-products. *Food Chem* 2011, 134:1296–1307.

9. Ates S, Mehmetoglu U: A new method for immobilization of galactosidase and its utilization in a plug flow reactor. *Process Biochem* 1997, 32:433–436.

10. Cheirsilp B, Jeamjounkhaw P, Kittikun AH: Optimizing an alginate immobilized lipase for monoacylglycerol production by the glycerolysis reaction. *J Mol Cat B: Enz* 2009, 59:206–211.

11. Ortega N, Perez-Mateos M, Pilar MC, Busto MD: Neutrase immobilization on alginate-glutaraldehyde beads by covalent attachment. *J Agric Food Chem* 2009, 57:109–115.

12. Danial EN, Elnashar MMM, Awad GEA: Immobilized inulinase on grafted alginate beads prepared by the one-step and the two-steps methods. *Ind Eng Chem Res* 2010, 49:3120–3125.

13. Leaes E, Zimmermann E, Souza M, Ramon A, Mezadri E, Dal Prá V, Terra L, Mazutti M: Ultrasound-assisted enzymatic hydrolysis of cassava waste to obtain fermentable sugars. *Bio Eng* 2013, 115:1–6.

14. Rocha JR, Catana R, Ferreira BS, Cabral JMS, Fernandes P: Design and characterization of an enzyme system from inulin hydrolysis. *Food Chem* 2006, 95:77–82.

15. Yewale T, Singhal RS, Vaidja AA: Immobilization of inulinase from *Aspergillus niger* NCIM 945 on chitosan and its application in continuous inulin hydrolysis. *Biocatal Agric Biotech* 2013, 2:96–101.

16. Richeti A, Munaretto CB, Lerin LA, Batistella L, Oliveira JV, Dallago RM, Astolfi V, Di Luccio M, Mazutti M, de Oliveira D, Treichel H: Immobilization of inulinase from Kluyveromyces marxianus NRRL Y-7571 using modified sodium alginate beads. *Bioprocess Biosyst Eng* 2012, 35:383–388.

Chapter 6

PARTICLE HANDLING TECHNIQUES IN MICROCHEMICAL PROCESSES

Brian S. Flowers and Ryan L. Hartman

Department of Chemical and Biological Engineering, The University of Alabama, Box 870203, Tuscaloosa, AL 35487-0203, USA

ABSTRACT

The manipulation of particulates in microfluidics is a challenge that continues to impact applications ranging from fine chemicals manufacturing to the materials and the life sciences. Heterogeneous operations carried out in microreactors involve high surface-to-volume characteristics that minimize the heat and mass transport resistances, offering precise control of the reaction conditions. Considerable advances have been made towards the engineering of techniques that control particles in microscale laminar flow, yet there remain tremendous opportunities for improvements in the area of chemical processing. Strategies that have been developed to successfully advance systems involving heterogeneous materials are reviewed and an outlook provided in the context of the challenges of continuous flow fine chemical processes.

INTRODUCTION

The field of microfluidics has grown extensively over the last decade, interfacing problems in biology, chemistry, and materials science to name a few [1,2,3,4,5,6,7,8,9,10]. A major challenge that continues to limit this exciting field is the ability to perform operations involving particles in microfluidic devices [11]. Heterogeneous systems of relevance to microfluidics are becoming increasingly common. A lack of technological approaches or fundamental understanding of how to deal with particulate matter in microscale laminar flow often leads to devices that clog, significantly reducing device life-cycles, or worse, rendering them inoperable. Advancing the ability to control particles in microfluidics [2,12,13,14,15,16,17,18,19,20,21,22,23,24,25,26,27,28,29,30,31,32,33,34,35,36,37,38,39,40,41] promises to create new areas of research and discovery with microsystems.

There are numerous microfluidic problems where solids are encountered yet limit the discovery of new science and engineering insights. Performing synthetic chemistry in microreactors can improve the selectivity and the yield while simultaneously elucidating the kinetic information needed to scale-up from the laboratory to an industrial process [42,43,44]. The majority of reactions relevant to the preparation of fine chemicals, however, involve the use of solid reagents, catalysts, products, and by-products. The manipulation of cells and biomolecules in microfluidics also demands control strategies for heterogeneous matter [6,20,45,46,47,48]. Microfluidic cell culturing can offer improvements on the nutrient conditions and hence cell yield, therapy, and response [49,50]. Devices have been engineered small enough to constrain a single cell, revealing deeper understanding of individual cellular mechanics [51,52]. Healthcare diagnostics and purification devices also take advantage of micro-scale flows to enhance the selectivity of biofluid separations, which can be used to isolate and purify proteins, cells, and lipids among others [53,54]. The microfluidic synthesis of inorganic crystals, another example of the flow and reaction of solids in micro-scales, can generate advanced nanomaterials for applications in catalysis, material science, and healthcare [5,10,55,56,57,58,59,60,61]. There exists the potential for device clogging in each of the aforementioned systems. Understanding the differences between the particle interaction mechanisms that constrain fluid flow is critical to exploiting microfluidics for research discovery and the development of devices that improve society.

There are important design considerations to be made when engineering microfluidic devices that will encounter particles. The mechanism(s) of device clogging depends on the nature of the particle-to-particle and particle-to-device interactions. Investing time up front to identify the dominant interactions can mitigate or prevent device failure. Techniques that can be applied to manage solids generally fall into one of two categories: active or passive particle manipulation. A combined understanding of the clogging mechanism(s) with the appropriate engineering strategy can ensure the flow of new knowledge.

This mini-review summarizes the fundamental concepts that one might consider when encountering solids in microfluidic systems, while the theme of chemical manufacturing is emphasized. Clogging mechanisms are described and examples drawn from the literature highlighted to demonstrate the basic concepts. A toolbox of engineering strategies to manage solids is presented, including advancements in both active and passive particle manipulation techniques. Finally, current limitations are discussed and opportunities for future advancements are delineated.

PARTICLES ENCOUNTERED IN MICROREACTORS

The synthetic methodologies applied in the preparation of pharmaceuticals and fine chemicals commonly involve solids in the form of reagents, catalysts, inorganic by-products, and organic products [62,63]. The coupling of aromatic molecules to prepare useful intermediates and final products presents virtually unlimited pathways and methodologies that involve the use of solids. Palladium-catalyzed bond forming reactions, for example, have found broad utility in the preparation of pharmaceuticals, natural products, and specialty materials [64,65,66]. This ubiquitous class of transition-metal catalyzed reactions commonly involves the use of starting materials that are introduced as solids into chemical reactors. As part of the catalytic cycle, palladium precipitates as a colloid (e.g., palladium black) and insoluble inorganic halide salts are formed as by-products. The example of palladium-catalyzed amination is one of many methodologies in pharmaceuticals and fine chemicals that involve substrates, catalyst, and by-products that become insoluble during chemical transformations.

Microfluidic devices also find utility in operations involving insoluble biomolecules, cells, and the synthesis of nanocrystals. The continuous separation of blood cells from lipids [19,20,22] or operations involving proteins [67,68] creates the potential for adhesion and/or agglomeration. The microfluidic synthesis of inorganic nanocrystals (e.g., CdSe nanoparticles) [58,61,69] has also been demonstrated to improve material properties, yet the continuous production creates the potential for reactor accumulation.

Microreactors can be used to accurately and rapidly extract the intrinsic kinetic information needed to scale-up microfluidic processes from the laboratory to intermediary and full production. Strategies are therefore needed to take advantage of microreactors for continuing discovery and development. Choosing the appropriate strategy depends on the chemistry and the physics of particle interactions that lead to clogging events.

PARTICLE FORMATION, STABILITY, AND ACCUMULATION CONSIDERATIONS

Nucleation Theory

The root of the clogging problem in microfluidics begins with events that take place at the molecular scale, while keeping in mind that much larger scales of tens-to-hundreds of microns ultimately result in device failures. If the thermodynamics predict reactive particle formation, then nucleation theory is the appropriate starting point to understand how to control particles in microfluidics [15,34,35,36,70]. The solubility product of each compound should be estimated for a given reaction solvent and the corresponding supersaturation ratio, the driving force for nucleation, approximated. Precipitation can be anticipated when supersaturation exists, where the primary rate of nucleation is dependent on the induction time; the time from when supersaturated conditions have been achieved to the appearance of the first detectable nucleus. It is common for nucleation kinetic predictions to vary by orders of magnitude due to heterogeneous nucleation and other molecular level variations. The presence of impurities, the reactor surface characteristics, axial and radial concentration and temperature gradients, and molecular interactions between compounds can all influence the prediction of the nucleation kinetics. If a reasonable estimate for the induction time has been experimentally validated, then a quantity useful in evaluating the potential for reactor fouling is the ratio of the induction time, t_{ind}, relative to the time necessary for a molecule to flow from the entrance to the exit of a reactor (i.e., the residence time τ), α, which can be expressed as [62]:

$$\alpha = \frac{t_{ind}}{\tau}$$

(1)

Such a dimensionless quantity is, as a worst case, useful for reducing the risk of particle accumulation on continuous reactor surfaces. In continuous operation, the large surface-to-volume ratios of microreactors are exposed to infinite residence time. Evaluation of Equation (1) can offer insight, from a heterogeneous nucleation perspective, on whether or not production goals can be met before any material accumulation begins. Classical nucleation theory and extensive models derived from it have been developed for inorganic and organic crystallization [71,72,73,74].

Particle Inertia

A reasonable starting point once bulk particles grow to a critical size (or in the absence of particle growth) is the consideration of inertial forces, which influence the motions of particles in microscale laminar flow. Stokes number, the ratio of the viscous to the inertial forces acting on a flowing particle, written as [75]

$$St = \frac{2}{9}\left(\frac{W}{D}\right)^2 \frac{\rho_p}{\rho_s} Re$$

(2)

can be estimated to approximate the potential for particle-to-wall interactions. When $St < 1$, particles do not spend enough time in the vicinity of a microchannel wall to undergo inertial impaction. When $St > 1$, particle trajectories along streamlines are modified and collisions with microreactor walls possible. When the inertial forces are favorable for impaction, or the onset of crystal growth on reactor surfaces takes place, estimation of the constriction rate, r_c, gives the ratio of the constriction rate to the convective transport rate through the reactor and in the axial direction, written as

$$\beta = \frac{r_c \tau}{L} \tag{3}$$

Equation (3) is useful to approximate the accumulation time scale relative to the residence time, which similar to Equation (1) can help to identify a process window in which the accumulation influences production. Comprehensive understanding of each materials class, crystalline or amorphous and organic or inorganic, is needed to most appropriately engineer continuous reactors that undergo steady-state operation while manufacturing microfluidic products.

Colloidal Attraction and Repulsion Physics

Colloidal attraction and repulsion theory describes the significance of sub-micron particle interaction energies and their role on aggregate formation and/or wall deposition. Consider two colloidal particles flowing along laminar streamlines in a microfluidic device (i.e., Reynolds number (Re) < 1000) and in the axial direction (i.e., the z-direction) according to Figure 1. Neglecting body forces such as gravity, an assumption that generally holds for sub-micron particles, approximates the settling time relative to the axial particle motion through the microfluidic device to be large. Thus, particle-to-particle, particle-to-wall, and hydrodynamic interactions represent the dominant forces acting on each particle, as illustrated in Figure 1.

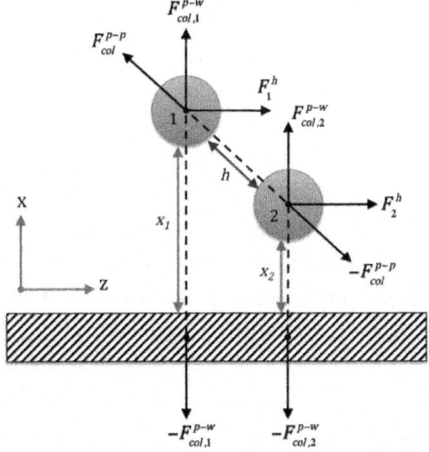

Figure 1. Colloidal and hydrodynamic forces acting on two spheres in laminar flow near the surface of a microfluidic device.

The total particle-to-particle colloidal force, $F_{col}{}^{p\text{-}p}$, is made up of three components with the first being van der Waals attraction for particles of identical surface chemistry, $V_A(h)$, [76,77]

$$F_{col}^{p-p}(h, a) = -\frac{d}{dh}[V_A(h, a) + V_R(h, a) + V_B(h, a)]$$

(4)

The particle-to-particle colloidal force is also a function of the electrostatic interaction energy, $V_R(h)$, [75] and the Born interaction energy, $V_B(h)$ [78]. Each of the three interaction energies are influenced by the interparticle distance, h, and the particle diameters, a. When particle-to-wall attraction or repulsion exists, quite possible in complex synthesis routes that often involve ionic species, the particle-to-wall colloidal force should be considered, $F_{col}{}^{p\text{-}w}$, expressed in terms of the electrostatic free energy (G_R) [79] and the free energy of adhesion (G_A) [80], by

$$F_{col}^{p-w}(x, a) = \left[-\frac{\kappa}{\varepsilon\pi}\left(\frac{e}{kT}\right)^2\frac{dG_R(x, a)}{dx}\right] + \left[-12\pi x_0^2\frac{dG_A(x, a)}{dx}\right]$$

(5)

where κ is the Debye parameter, ε the permittivity, k Boltzmann constant, e the electronic charge, T the temperature, and x_0 the minimum equilibrium distance between the particle and the wall. Here, note that the particle-to-wall separation distance, x, influences the magnitude of the force, and thus only particles flowing near the reactor wall are expected to encounter such an energy field. Although our example is a simplified scenario, estimation of the magnitudes of the particle-to-particle and the particle-to-wall interaction forces yields important information for the design of continuous reactor operating conditions. As an example, the ratio of the interparticle and the particle-to-wall colloidal attraction magnitudes,

$$\lambda = \frac{\left|F_{col}^{p-p}\right|}{\left|F_{col}^{p-w}\right|}$$

(6)

offers mechanistic understanding of the potential solids handling challenges. When $\lambda > 1$, aggregation may take place, leading to microscopic blockages. When $\lambda < 1$, there exists the potential for the accumulation of particles on surfaces. In a system of particles, Equation (6) can be expressed as the sum of the particle-to-particle and particle-to-wall attractive interactions. A force balance in equilibrium and without gravitational forces can be rearranged as

$$\eta = \frac{F^h}{\left(F_{col}^{p-p} + F_{col}^{p-w}\right)}$$

(7)

One observes that $\eta < -1$ when the net hydrodynamic force acting on a particle exceeds the colloidal forces. For $\eta > -1$ values, low Reynolds number flow for instance, colloidal particle forces can generate agglomerates or accumulate on microreactor surfaces. The relationship of Equation (7), however, yields a deeper understanding when a comparison of the important time scales is made. When $\alpha < 1$, defining the work time for each force reveals the expression,

$$\omega = \frac{|F^h| L\tau}{\left(|F_{col}^{p-p}| h t_h + |F_{col}^{p-w}| x t_x\right)}$$

(8)

The hydrodynamic work time relative to the colloidal attraction work time, ω, only considering particle attraction, describes the probability of either aggregation or wall accumulation in terms of the overall residence time. For example, when $\lambda \sim 1$ and $t_x \gg t_h$ (or the distance $x \gg h$), then $\omega_w = (|F^h| L\tau)/ (|F_{col}^{p-w}| x t_x)$ and ω_w values < 1 imply that particles are transported axially throughout a microreactor before accumulation has had enough time to occur. Similarly, when $\lambda \sim 1$ and $t_h \gg t_x$ (or the distance $h \gg x$), then $\omega_p = (|F^h| L\tau)/ (|F_{col}^{p-p}| h t_h)$ and ω_p values < 1 imply that particles are transported axially throughout a microreactor before aggregation has had enough time to occur. Such a relationship is critical in understanding when particle aggregates that remain in the bulk flow exit a microreactor in time to eliminate any pressure losses. Additional theoretical models relating particle aggregation physics can be applied to predict the growth rates for a population of suspended particles.

Bridging, Constriction, and Random Detachment

Three fundamentally different mechanisms, (1) bridging, (2) constriction, or (3) random detachment, can result in pressure losses during the micro-to-meso-scale laminar flow of particulates. A combination of any of the three mechanisms can propagate when chemical reactions occur. Each of the three mechanisms may also arise in the absence of particle formation or growth.

Particle motions tracking lamellae are possible when the net hydrodynamic forces are greater than the colloidal and inertial forces (i.e., $\eta < -1$, $\omega < 1$, and $St < 1$). A reduction in the cross-sectional flow path under such conditions can lead to the simultaneous arrival of particles at the constriction point, and in turn generate bridging events; a statistical or velocity controlled phenomenon depending on the operating regime [76,81]. Bridging can also occur in the absence of reduced cross-sections. As shown in Figure 2a, bridging constrains fluid flow (not considering attraction or repulsion) for channel width-to-particle size aspect ratios, $D/a < 10$. The bridging, also shown in the microreactor image of Figure 2b for NaCl formation, is imminent when $St > 1$, $\eta > -1$, and

$\omega > 1$. Under such conditions particles traveling near the reactor wall deposit, forming dendrites that eventually bridge.

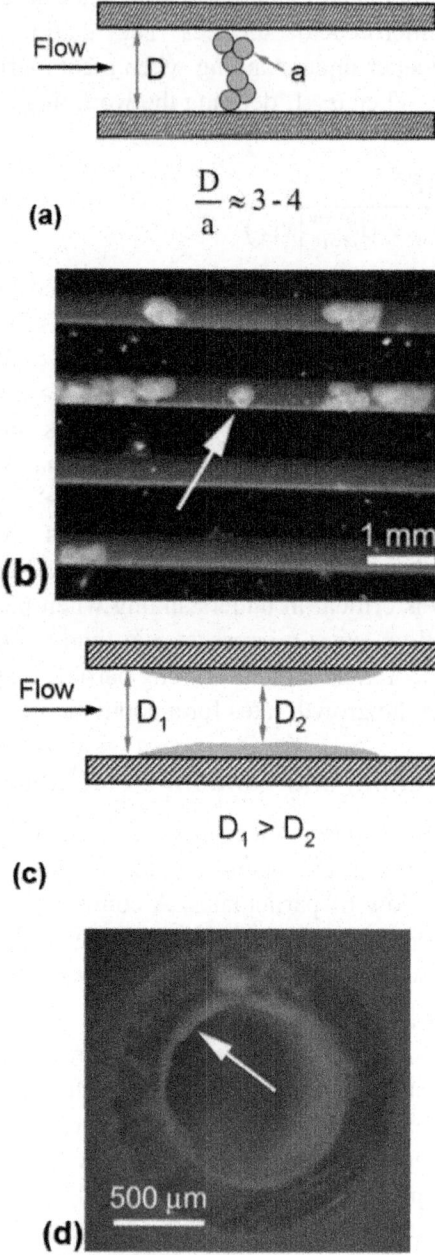

Figure 2. Example plugging mechanisms that occur in micro-scale laminar flow. **(a)** Particle bridging constrains fluid flow when the aspect ratio of the channel width to

the particle sizes is about 3–7. (**b**) The bridging of NaCl crystals in a microreactor has been observed to occur while performing palladium-catalyzed bond forming reactions. (**c**) The microreactor cross-section is reduced, or constricted, when wall deposition or nucleation occurs, which also has been observed during (**d**) reactive NaCl precipitation in fluoropolymer capillaries. (Reprinted with permission from Reference [35] Copyright 2010 American Chemical Society).

Fluid flow may also be constrained when microreactor cross-sections diminish with time, [82] illustrated in Figure 2(c,d). The constriction occurs when particles deposit on surfaces or when nucleation followed by growth takes place. In either scenario, wall build-ups present a challenge to continuous production in microfluidic systems because process shutdowns are needed to remediate accumulated material. The random detachment of the accumulated material, another potential challenge, is virtually unpredictable yet can generate macroscopic blockages that severely limit fluid flow.

TECHNIQUES TO CONTROL PARTICLE TRANSPORT

When particle-to-particle or particle-to-wall attraction exists, nano-to-micro-scale phenomena, the bridging or constrictions of microfluidic flow paths are imminent. The nucleation and growth of clusters in the bulk or on device surfaces, a molecular scale problem, can result in the bridging or the constriction as well. Either mechanism, from the molecular to the micro-scale, requires the conception of engineering techniques to control particle transport. Such techniques generally fit into one of two categories: (1) passive particle manipulation or (2) active particle manipulation.

Passive Particle Manipulation

Passive techniques are those that do not impose any external forces beyond the steady-state energy required to commence and sustain fluid flow. Hydrodynamic flow focusing, an example of passive particle manipulation, has been demonstrated as an effective technique to order micron-sized particles in microfluidics. The streaming of particle suspensions (e.g., polystyrene microspheres) is made possible via flow through spiral geometries that establish Dean flow [83]. The existence of hydrodynamic force vectors tangential to the flow cross-section enables centerline particle equilibrium (i.e., the ratio of the inertial lift to the drag force magnitudes is optimized), and thus the longitudinal ordering and lateral flow focusing of random particles is possible [83]. The Dean number, expressed as

$$De = Re\sqrt{\frac{D}{2R}}$$

(9)

estimates the existence of cross-sectional velocity components where R is the radius of microchannel curvature. In straight microchannels, De = 0, indicating the absence of a drag force due to the transverse Dean flows [83]. The magnitude of the transverse Dean flow increases, as seen in Equation (9), with a decreasing radius of curvature, and thus the microchannel geometry can be microfabricated to spatially position particles of different momentums [47]. A challenge with passive flow focusing, however, is its ability to manage particle concentrations encountered in fine chemical production when particle-to-particle and particle-to-wall attraction exists (i.e., $\eta > -1$, and $\omega > 1$). Furthermore, particle momentum can lead to migration across streamlines, creating particle-to-particle and particle-to-wall collisions.

The isolation of particles that tend to agglomerate or stick to surfaces has been accomplished by the encapsulation of reactions within dispersed droplets, as highlighted in Figure 3(a,b). Protein crystals (Figure 3a), [67] or other insoluble compounds such as polymers (Figure 3b), [38,39] are limited in their interaction with microchannel surfaces in the presence of an immiscible barrier between the phases. Interestingly, immiscible gas-liquid interfaces [84,85,86] have been demonstrated as possible separation techniques; micron-sized particles accumulate at the lower surface-free-energy interfaces (see Figure 3c) [87,88,89]. In a system of excess particles, shown in Figure 3d, armored bubbles form when liquid-liquid immiscible interfaces are occupied entirely by particulates [90,91]. Establishing the equilibrium illustrated in Figure 3d is possible when the colloidal forces are stabilized, which depends on the ionic strength of the liquid media; an important design consideration and a challenge to control in complex heterogeneous reaction pathways exposed to convective forces.

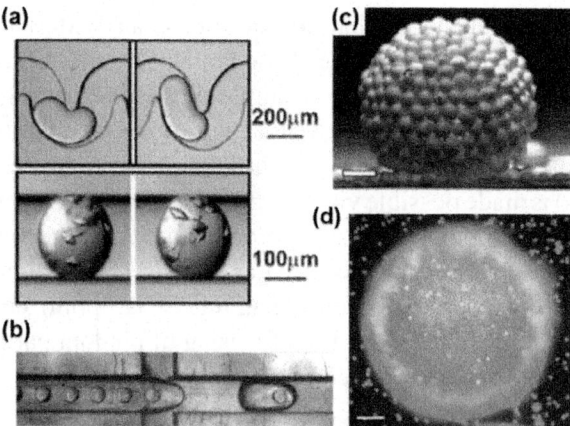

Figure 3. Examples of passive techniques engineered to control particles in microscale laminar flow. Immiscible liquid-liquid flows enable the encapsulation of solids within

dispersed droplets and thereby limit the particle-to-wall interactions, shown for (**a**) protein crystals (Reprinted with permission from Reference [67] Copyright 2005 American Chemical Society) and (**b**) polymers (Copyright Wiley-VCH Verlag GmbH & Co. KGaA, reproduced with permission) [39]. Immiscible interfaces also provide lower surface free energy sites for particles to accumulate, thus (**c**) collecting particles at gas-liquid interfaces (scale bar = 400 μm) (Reprinted with permission from Reference [88] Copyright 2006 American Chemical Society)and/or (**d**) creating armored bubbles is possible (scale bar = 60 μm) (Reprinted by permission from Macmillan Publishers Ltd: Nature Materials [90], copyright 2005).

Active Particle Manipulation

Active particle techniques involve external forces beyond the steady-state energy required to commence and sustain fluid flow. Pulse flow could be considered an active particle manipulation technique and has been exploited to engineer dry powder injections within microfluidic devices [31]. For D/a aspect ratios that do not bridge, inert gas has been used to carry dry pharmaceutical powders within microscale devices [31]. The working principle allows the formation of a fluidized particle bed in microfluidics when particle-to-particle attractive interactions do not exist [31].

The use of light and electron gradients also represents active particle manipulation techniques. For example, optical particle displacement has been accomplished by generating bubbles with sufficiently intense laser light [92]. Single and tandem bubbles generated by infrared and green lasers displaced particles such as polystyrene microspheres [92]. As shown in Figure 4a, particles were displaced tens-of-microns and within microseconds [92]. Optical force switching, another active technique utilizing light, has been engineered for the microfluidic separation of mammalian cells [93]. The diminishing light intensity with increasing cross-sectional dimensions presents a challenge, in general, for scaling up optical particle control techniques, especially when considering fine chemical manufacturing. Dielectrophoretic particle separations and cell isolations are also possible when an electric field is imposed upon a system of charged particles traveling through a microfluidic device. The governing principles that establish a force balance for the steady-state separation of particles have previously been described [41,94]. As seen in the schematic of Figure 4b, the fabrication of devices with deflection electrodes enables particle sorting where the threshold velocity for the displacement of different sized latex microspheres scales with the voltage amplitude (Figure 4c) [94].

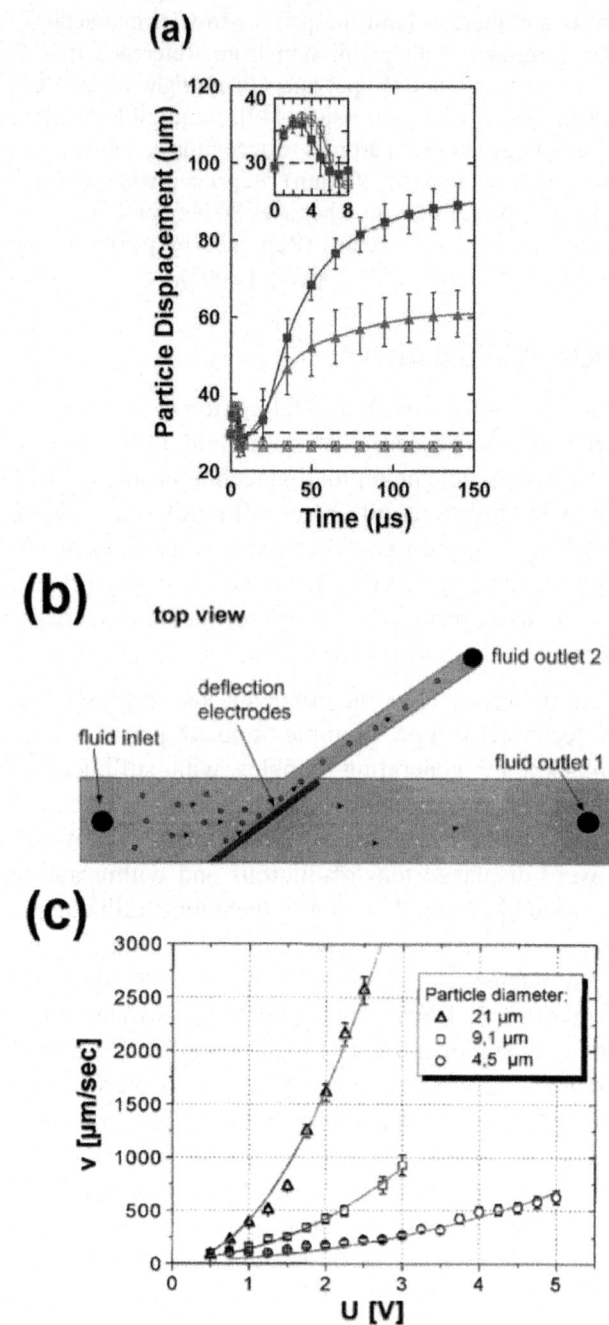

Figure 4. (a) Laser light driven particle displacements made possible through the generation of single and tandem bubbles (Reprinted with permission from [92]. Copyright

2010, American Institute of Physics). (b) Dielectrophoretic particle sorting by the microfabrication of deflection electrodes. (c) The threshold velocity for the displacement of different sized latex microspheres scales with the voltage amplitude. (Copyright Wiley-VCH Verlag GmbH & Co. KGaA, reproduced with permission) [94].

Magnetism, another active technique, also enables the separation of particulates in microfluidics [2,24,30,32,33,95,96]. When particles flow by a magnetic field (see Figure 5a), the magnetically induced velocity vector, u_{mag}, is described by the magnetic force, F_{mag}, exerted on the particle relative to the viscous drag force, F_{vis}, by [30]

$$u_{mag} = \frac{F_{mag}}{F_{vis}} = \frac{\Delta\chi V_p(\nabla \cdot B)B}{6\pi\mu a K} \tag{10}$$

where $\Delta\chi$ is the magnetic susceptibility difference between the particle and the liquid medium, V_p the particle volume, B the externally applied magnetic flux density, ∇B the flux density gradient, μ the liquid viscosity, and K the permeability of a vacuum [30]. One can infer from Equation (10) that magnetically driven particle motions occur when the magnitude of the magnetic force exceeds the viscous force. Equation (10) additionally illustrates the critical dependence of magnetically driven particle separations on the particle radius, a, and the magnetic characteristics of the particle (i.e., $\Delta\chi$ by χ_p). There exists potential for magnetic separations in fine chemical and some inorganic crystal manufacturing. Aggregates (i.e., agglomerates) behave as larger particle sizes in a magnetic field, and thus the continuous, laminar separation of the aggregates is possible [30]. In practice, a distribution of particle sizes (superparamagnetic particles, e.g., Dynabeads), shown in Figure 5b, were deflected each corresponding to the parallel outlets of Figure 5b [30]. Scaling up magnetic particle separations for the continuous flow manufacturing of many microfluidic processes remains a challenge as the viscous force magnitude increases with Re and the delivery of adequate magnetic fields, a capital cost consideration, requires design innovations.

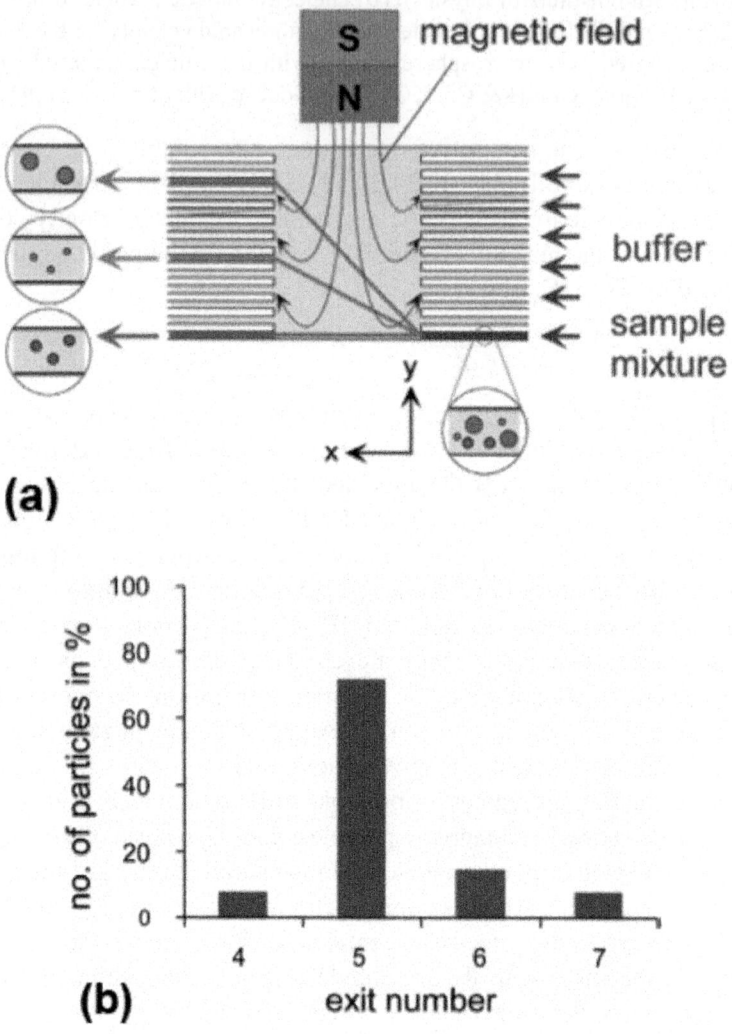

Figure 5. (a) Microfluidic separation of superparamagnetic particles in the presence of a magnetic field. (b) The collection of particles exiting the device enables the isolation of the fraction of particles deflected by magnetism.(Reprinted with permission from Reference [30] Copyright 2004 American Chemical Society).

It is also readily known that acoustic waves displace particles traveling along lamellae [18,20,21,22,23,25,26,27,97,98,99,100,101,27,97]. High frequency (e.g., MHz) acoustic standing waves force particles to pressure nodes in laminar flow, which has been demonstrated for the separation of red blood cells from lipids [23,25,26,102]. A particle traveling through an acoustic standing wave experiences a primary force, F_a^1, described by

$$F_a^1 = \left(\frac{\pi p_0^2 V_c \beta_s}{2\lambda_a}\right) \cdot \phi(\beta, \rho) \cdot sin(2kx) \tag{11}$$

where p_0 is the acoustic pressure amplitude, V_c the particle volume, λ_a the acoustic wavelength, k the wavenumber, and x the distance from a pressure node. The primary force is considered significant when a phase difference parameter, φ, expressed as

$$\phi = \frac{5\rho_p - 2\rho_s}{2\rho_p + \rho_s} - \frac{\beta_p}{\beta_s} \tag{12}$$

is non-zero [23,25,26,102]. The terms ρ_i and β_i are the density and compressibility of the solvent (s) and the particle (p), respectively. When particle diameters are less than half the wavelength, the acoustic streaming of suspended particles is possible on-chip, shown in Figure 6a, by integrating ultrasonic transducers with microfluidic devices. Secondary forces, also significant when a concentrated system of particles scatters acoustic standing waves, have been described in a more detailed review [22].

Acoustic waveforms in the range of kHz, for example ultrasonic cleaning equipment, have been demonstrated as effective in carrying out microchemical transformations that generate solids [15,35,64]. On-chip ultrasound has been engineered to manage the NaCl formation during palladium-catalyzed bond forming reactions [13]. As an example, the by-product MnO_2 formed during the Nef oxidation leads to flow-induced clogging, which can be prevented using ultrasound (shown in Figure 6b) [103]. Overall, the integration of acoustics with microreactors has tremendous potential that fosters a need for scale-up innovations.

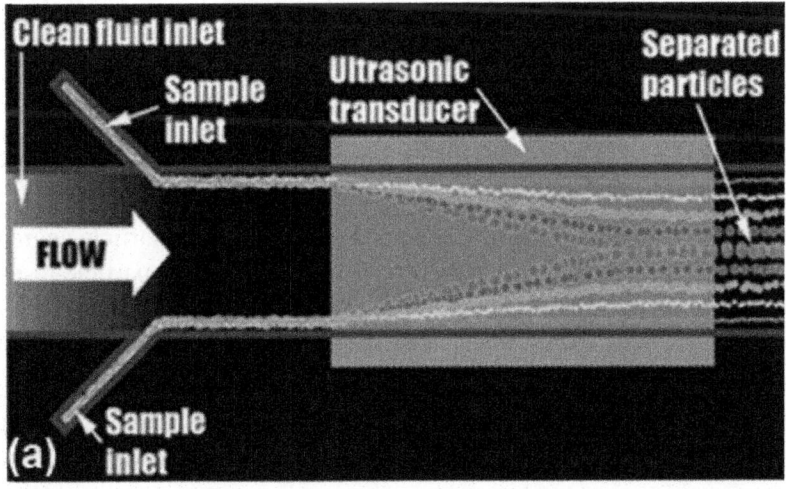

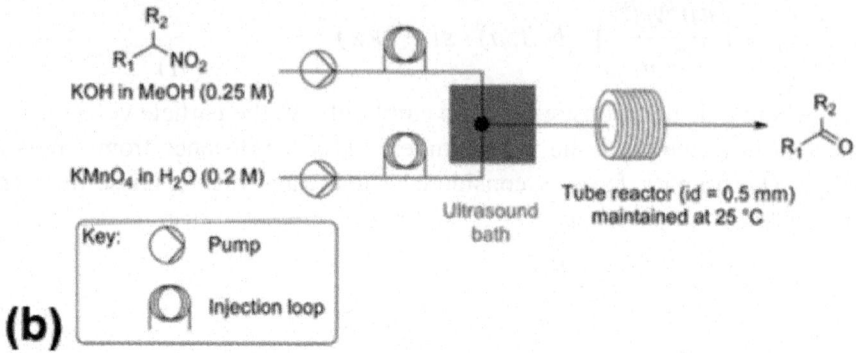

Figure 6. (a) The separation of microparticles via acoustic streaming in microfluidics (Reprinted with permission from [25]. Copyright 2007 American Chemical Society). **(b)** Performing solids forming reactions, such as the nef oxidation, in the presence of ultrasound has been shown to mitigate flow-induced clogging (Reprinted with permission from [103]. Copyright 2010 American Chemical Society).

Advancements continue to be made towards the innovation of new techniques for solids handling in microreactors. Continuous-stirred tank reactors and mechanically agitated flow cells [104] have been demonstrated as effective methods to carry out particle forming reactions. Nevertheless, the design of solids handling reactors only represents part of the challenge; an entire chemical process is made up of many different unit operations each where solids in continuous flow could be problematic.

OUTLOOK AND CHALLENGES

The thermodynamic and the kinetic conditions that make solids formation probable necessitate understanding of the relative time scales in continuous flow microreactors. Such understanding is also needed when particles undergo laminar flow through microfluidic devices. Engineering techniques to control particles, whether generated or not, requires the consideration of particle-to-particle and particle-to-wall attraction and repulsion. Overcoming particle-to-wall attraction that results in accumulation is possible through the evaluation of the force magnitudes acting on each particle. Minimizing particle aggregation leading to macroscopic blockages is possible by assessing particle-to-particle attraction in relationship to the residence time throughout a microreactor. The analogy is particularly useful in the design of continuous microreactors for chemical processes.

A number of creative passive and active techniques have been engineered for the manipulation of particles in microscale laminar flow, which have the potential to impact chemical processing. The integration of such techniques, however, is not without challenges. The majority of passive techniques do not solve the problem of particle aggregation or wall accumulation due to attraction. Similarly, scaling up active techniques introduces design challenges in terms of the imposed forces relative to the viscous and inertial forces. There remains tremendous opportunity for new approaches, building on existing solids handling strategies, to ensure that continuous microfluidic processes do not require remediation.

REFERENCES

1. Ottino, J.M.; Wiggins, S. Introduction: Mixing in microfluidics. Philos. Trans. R. Soc. A **2004**, 362, 923–935.

2. Pamme, N. Magnetism and microfluidics. Lab Chip **2006**, 6, 24–38.

3. Stone, H.A.; Stroock, A.D.; Ajdari, A. Engineering flows in small devices: Microfluidics toward a lab-on-a-chip. Ann. Rev. Fluid Mech. **2004**, 36, 381–411.

4. Whitesides, G.M. The origins and the future of microfluidics. Nature **2006**, 442, 368–373.

5. Marre, S.; Jensen, K.F. Synthesis of micro and nanostructures in microfluidic systems. Chem. Soc. Rev. **2010**, 39, 1183–1202.

6. El-Ali, J.; Sorger, P.K.; Jensen, K.F. Cells on chips. Nature **2006**, 442, 403–411.

7. Hartman, R.L.; Jensen, K.F. Microchemical systems for continuous-flow synthesis. Lab Chip **2009**, 9, 2495–2507.

8. Jensen, K.F. Microreaction engineering—Is small better? Chem. Eng. Sci. **2001**, 56, 293–303.

9. Atencia, J.; Beebe, D.J. Controlled microfluidic interfaces. Nature **2005**, 437, 648–655.

10. Krishnadasan, S.; Brown, R.J.C.; Demello, A.J.; Demello, J.C. Intelligent routes to the controlled synthesis of nanoparticles. Lab Chip **2007**, 7, 1434–1441.

11. Mukhopadhyay, R. When microfluidic devices go bad—How does fouling occur in microfluidic devices, and what can be done about it? Anal. Chem. **2005**, 77, 429–432.

12. Horie, T.; Sumino, M.; Tanaka, T.; Matsushita, Y.; Ichimura, T.; Yoshida, J. Photodimerization of maleic anhydride in a microreactor without clogging. Org. Proc. Res. Dev. **2010**, 14, 405–410.

13. Kuhn, S.; Noel, T.; Gu, L.; Heider, P.L.; Jensen, K.F. A Teflon microreactor with integrated piezoelectric actuator to handle solid forming reactions. Lab Chip 2011, 11, 2488–2492.

14. Nagasawa, H.; Mae, K. Development of a new microreactor based on annular microsegments for fine particle production. Ind. Eng. Chem. Res. 2006, 45, 2179–2186.

15. Noel, T.; Naber, J.R.; Hartman, R.L.; McMullen, J.P.; Jensen, K.F.; Buchwald, S.L. Palladium-catalyzed amination reactions in flow: Overcoming the challenges of clogging via acoustic irradiation. Chem. Sci. 2011, 2, 287–290.

16. Poe, S.L.; Cummings, M.A.; Haaf, M.R.; McQuade, D.T. Solving the clogging problem: Precipitate-forming reactions in flow. Angew. Chem. Int. Ed. 2006, 45, 1544–1548.

17. Pribyl, M.; Snita, D.; Marek, M. Nonlinear phenomena and qualitative evaluation of risk of clogging in a capillary microreactor under, imposed electric field. Chem. Eng. J. 2005, 105, 99–109.

18. Bengtsson, M.; Laurell, T. Ultrasonic agitation in microchannels. Anal. Bioanal. Chem. 2004, 378, 1716–1721.

19. Evander, M.; Johansson, L.; Lilliehorn, T.; Piskur, J.; Lindvall, M.; Johansson, S.; Almqvist, M.; Laurell, T.; Nilsson, J. Noninvasive acoustic cell trapping in a microfluidic perfusion system for online bioassays. Anal. Chem. 2007, 79, 2984–2991.

20. Hawkes, J.J.; Barber, R.W.; Emerson, D.R.; Coakley, W.T. Continuous cell washing and mixing driven by an ultrasound standing wave within a microfluidic channel. Lab Chip 2004, 4, 446–452.

21. Hawkes, J.J.; Coakley, W.T. Force field particle filter, combining ultrasound standing waves and laminar flow. Sens. Act. B Chem. 2001, 75, 213–222.

22. Laurell, T.; Petersson, F.; Nilsson, A. Chip integrated strategies for acoustic separation and manipulation of cells and particles. Chem. Soc. Rev. 2007, 36, 492–506.

23. Nilsson, A.; Petersson, F.; Jonsson, H.; Laurell, T. Acoustic control of suspended particles in micro fluidic chips. Lab Chip 2004, 4, 131–135.

24. Pamme, N. Continuous flow separations in microfluidic devices. Lab Chip 2007, 7, 1644–1659.

25. Petersson, F.; Aberg, L.; Sward-Nilsson, A.M.; Laurell, T. Free flow acoustophoresis: Microfluidic-based mode of particle and cell separation. Anal. Chem. 2007, 79, 5117–5123.

26. Petersson, F.; Nilsson, A.; Holm, C.; Jonsson, H.; Laurell, T. Continuous separation of lipid particles from erythrocytes by means of laminar flow and acoustic standing wave forces. Lab Chip **2005**, 5, 20–22.

27. Petersson, F.; Nilsson, A.; Jonsson, H.; Laurell, T. Carrier medium exchange through ultrasonic particle switching in microfluidic channels. Anal. Chem. **2005**, 77, 1216–1221.

28. Dittrich, P.S.; Tachikawa, K.; Manz, A. Micro total analysis systems. Latest advancements and trends. Anal. Chem.**2006**, 78, 3887–3907.

29. Pamme, N.; Koyama, R.; Manz, A. Counting and sizing of particles and particle agglomerates in a microfluidic device using laser light scattering: Application to a particle-enhanced immunoassay. Lab Chip **2003**, 3, 187–192.

30. Pamme, N.; Manz, A. On-chip free-flow magnetophoresis: Continuous flow separation of magnetic particles and agglomerates. Anal. Chem. **2004**, 76, 7250–7256.

31. Vilkner, T.; Shivji, A.; Manz, A. Dry powder injection on chip. Lab Chip **2005**, 5, 140–145.

32. Pamme, N.; Wilhelm, C. Continuous sorting of magnetic cells via on-chip free-flow magnetophoresis. Lab Chip **2006**, 6, 974–980.

33. Rodriguez-Villarreal, A.I.; Tarn, M.D.; Madden, L.A.; Lutz, J.B.; Greenman, J.; Samitier, J.; Pamme, N. Flow focussing of particles and cells based on their intrinsic properties using a simple diamagnetic repulsion setup. Lab Chip **2011**, 11, 1240–1248.

34. Hartman, R.L. Managing solids in microreactors for the upstream continuous processing of fine chemicals. Org. Proc. Res. Dev. **2012**.

35. Hartman, R.L.; Naber, J.R.; Zaborenko, N.; Buchwald, S.L.; Jensen, K.F. Overcoming the challenges of solid bridging and constriction during Pd-catalyzed C-N bond formation in microreactors. Org. Proc. Res. Dev. **2010**, 14, 1347–1357.

36. Kuhn, S.; Hartman, R.L.; Sultana, M.; Nagy, K.D.; Marre, S.; Jensen, K.F. Teflon-coated silicon microreactors: Impact on segmented liquid-liquid multiphase flows. Langmuir **2011**, 27, 6519–6527.

37. Honda, T.; Miyazaki, M.; Nakamura, H.; Maeda, H. Controllable polymerization of N-carboxy anhydrides in a microreaction system. Lab Chip **2005**, 5, 812–818.

38. Li, W.; Pharn, H.H.; Nie, Z.; MacDonald, B.; Guenther, A.; Kumacheva, E. Multi-step microfluidic polymerization reactions conducted in

droplets: The internal trigger approach. J. Am. Chem. Soc. **2008**, 130, 9935–9941.

39. Marcati, A.; Serra, C.; Bouquey, M.; Prat, L. Handling of polymer particles in microchannels. Chem. Eng. Tech. **2010**, 33, 1779–1787.

40. Yamada, M.; Seki, M. Hydrodynamic filtration for on-chip particle concentration and classification utilizing microfluidics. Lab Chip **2005**, 5, 1233–1239.

41. Kralj, J.G.; Lis, M.T.W.; Schmidt, M.A.; Jensen, K.F. Continuous dielectrophoretic size-based particle sorting. Anal. Chem. **2006**, 78, 5019–5025.

42. Hessel, V. Novel process windows—Gate to maximizing process intensification via flow chemistry. Chem. Eng. Technol. **2009**, 32, 1655–1681.

43. Kockmann, N.; Gottsponer, M.; Roberge, D.M. Scale-up concept of single-channel microreactors from process development to industrial production. Chem. Eng. J. **2011**, 167, 718–726.

44. Roberge, D.M.; Zimmermann, B.; Rainone, F.; Gottsponer, M.; Eyholzer, M.; Kockmann, N. Microreactor technology and continuous processes in the fine chemical and pharmaceutical industry: Is the revolution underway? Org. Proc. Res. Dev. **2008**, 12, 905–910.

45. Adamo, A.; Jensen, K.F. Microfluidic based single cell microinjection. Lab Chip **2008**, 8, 1258–1261.

46. Huh, D.; Gu, W.; Kamotani, Y.; Grotberg, J.B.; Takayama, S. Microfluidics for flow cytometric analysis of cells and particles. Physiol. Meas. **2005**, 26, R73–R98.

47. Kuntaegowdanahalli, S.S.; Bhagat, A.A.S.; Kumar, G.; Papautsky, I. Inertial microfluidics for continuous particle separation in spiral microchannels. Lab Chip **2009**, 9, 2973–2980.

48. Salieb-Beugelaar, G.B.; Simone, G.; Arora, A.; Philippi, A.; Manz, A. Latest developments in microfluidic cell biology and analysis systems. Anal. Chem. **2010**, 82, 4848–4864.

49. Bruzewicz, D.A.; McGuigan, A.P.; Whitesides, G.M. Fabrication of a modular tissue construct in a microfluidic chip.Lab Chip **2008**, 8, 663–671.

50. Marcy, Y.; Ishoey, T.; Lasken, R.S.; Stockwell, T.B.; Walenz, B.P.; Halpern, A.L.; Beeson, K.Y.; Goldberg, S.M.D.; Quake, S.R. Nanoliter reactors improve multiple displacement amplification of genomes from single cells. PLoS Genet.**2007**, 3, 1702–1708.

51. Kortmann, H.; Blank, L.M.; Schmid, A. Single Cell Analytics: An Overview. In High Resolution Microbial Single Cell Analytics; Muller, S., Bley, T., Eds.; Springer-Verlag Berlin: Berlin, Germany, 2011; Volume 124, pp. 99–122.

52. Schmid, A.; Kortmann, H.; Dittrich, P.S.; Blank, L.M. Chemical and biological single cell analysis. Curr. Opin. Biotechnol. **2010**, 21, 12–20.

53. Jaggi, R.D.; Sandoz, R.; Effenhauser, C.S. Microfluidic depletion of red blood cells from whole blood in high-aspect-ratio microchannels. Microfluid. Nanofluid. **2007**, 3, 47–53.

54. Pommer, M.S.; Zhang, Y.T.; Keerthi, N.; Chen, D.; Thomson, J.A.; Meinhart, C.D.; Soh, H.T. Dielectrophoretic separation of platelets from diluted whole blood in microfluidic channels. Electrophoresis **2008**, 29, 1213–1218.

55. Chabert, M.; Viovy, J.L. Microfluidic high-throughput encapsulation and hydrodynamic self-sorting of single cells.Proc. Natl. Acad. Sci. USA **2008**, 105, 3191–3196.

56. Dorvee, J.R.; Sailor, M.J.; Miskelly, G.M. Digital microfluidics and delivery of molecular payloads with magnetic porous silicon chaperones. Dalton Trans. **2008**, 721–730.

57. Hung, L.H.; Choi, K.M.; Tseng, W.Y.; Tan, Y.C.; Shea, K.J.; Lee, A.P. Alternating droplet generation and controlled dynamic droplet fusion in microfluidic device for CdS nanoparticle synthesis. Lab Chip **2006**, 6, 174–178.

58. Nakamura, H.; Yamaguchi, Y.; Miyazaki, M.; Maeda, H.; Uehara, M.; Mulvaney, P. Preparation of CdSe nanocrystals in a micro-flow-reactor. Chem. Commun. **2002**, 2844–2845.

59. Edel, J.B.; Fortt, R.; de Mello, J.C.; de Mello, A.J. Microfluidic routes to the controlled production of nanoparticles.Chem. Commun. **2002**, 1136–1137.

60. Khan, S.A.; Jensen, K.F. Microfluidic synthesis of titania shells on colloidal silica. Adv. Mat. **2007**, 19, 2556–2560.

61. Yen, B.K.H.; Gunther, A.; Schmidt, M.A.; Jensen, K.F.; Bawendi, M.G. A microfabricated gas-liquid segmented flow reactor for high-temperature synthesis: The case of CdSe quantum dots. Angew. Chem. Int. Ed. **2005**, 44, 5447–5451.

62. Hartman, R.L. Managing solids in microreactors for the upstream continuous processing of fine chemicals. Org. Proc. Res. Dev. **2012**.

63. Roberge, D.M.; Ducry, L.; Bieler, N.; Cretton, P.; Zimmermann, B. Microreactor technology: A revolution for the fine chemical and pharmaceutical industries? Chem. Eng. Technol. **2005**, 28, 318–323.

64. Noel, T.; Buchwald, S.L. Cross-coupling in flow. Chem. Soc. Rev. **2011**, 40, 5010–5029.

65. Surry, D.S.; Buchwald, S.L. Biaryl phosphane ligands in palladium-catalyzed amination. Angew. Chem. Int. Ed. **2008**,47, 6338–6361.

66. Suzuki, A. Cross-coupling reactions of organoboranes: An easy way to construct C–C bonds (Nobel Lecture). Angew. Chem. Int. Ed. **2011**, 50, 6722–6737.

67. Chen, D.L.; Gerdts, C.J.; Ismagilov, R.F. Using microfluidics to observe the effect of mixing on nucleation of protein crystals. J. Am. Chem. Soc. **2005**, 127, 9672–9673.

68. Koc, Y.; de Mello, A.J.; McHale, G.; Newton, M.I.; Roach, P.; Shirtcliffe, N.J. Nano-scale superhydrophobicity: Suppression of protein adsorption and promotion of flow-induced detachment. Lab Chip **2008**, 8, 582–586.

69. Marre, S.; Park, J.; Rempel, J.; Guan, J.; Bawendi, M.G.; Jensen, K.F. Supercritical continuous-microflow synthesis of narrow size distribution quantum dots. Adv. Mater. **2008**, 20, 4830–4834.

70. Kockmann, N.; Kastner, J.; Woias, P. Reactive particle precipitation in liquid microchannel flow. Chem. Eng. J. **2008**,135, S110–S116.

71. Mullin, J.W. Crystallization, 3rd ed; Butterworth Heinemann: Oxford, UK, 1997.

72. Slaughter, D.W.; Doherty, M.F. Calculation of solid-liquid equilibrium and crystallization paths for melt crystallization processes. Chem. Eng. Sci. **1995**, 50, 1679–1694.

73. Winn, D.; Doherty, M.F. Modeling crystal shapes of organic materials grown from solution. Aiche J. **2000**, 46, 1348–1367.

74. Zhang, Y.C.; Sizemore, J.P.; Doherty, M.F. Shape evolution of 3-dimensional faceted crystals. Aiche J. **2006**, 52, 1906–1915.

75. Russel, W.B.; Saville, D.A.; Schowalter, W.R. Colloidal Dispersions; Cambridge University Press: Cambridge, UK, 1992.

76. Ramachandran, V.; Fogler, H.S. Plugging by hydrodynamic bridging during flow of stable colloidal particles within cylindrical pores. J. Fluid Mech. **1999**, 385, 129–156.

77. Schenkel, J.H.; Kitchener, J.A. A test of the Derjaguin-Verwey-Overbeek theory with a colloidal suspension. Trans. Faraday Soc. **1960**, 56, 161–173.

78. Feke, D.L.; Prabhu, N.D.; Mann, J.A.; Mann, J.A. A formulation of the short-range repulsion between spherical colloidal particles. J. Phys. Chem. **1984**, 88, 5735–5739.

79. Bowen, R.W.; Sharif, A.O. Adaptive finite-element solution of the nonlinear Poisson-Boltzmann equation: A charged spherical particle at various distances from a charged cylindrical pore in a charged planar surface. J. Colloid Interface Sci. **1997**, 187, 363–374.

80. Bhattacharjee, S.; Sharma, A. Lifshitz-van der waals energy of spherical-particles in cylindrical pores. J. Colloid Interface Sci. **1995**, 171, 288–296.

81. Wyss, H.M.; Blair, D.L.; Morris, J.F.; Stone, H.A.; Weitz, D.A. Mechanism for clogging of microchannels. Phys. Rev. E Stat. Nonlin. Soft. Matter Phys. **2006**, 74.

82. Ramachandran, V.; Fogler, H.S. Multilayer deposition of stable colloidal particles during flow within cylindrical pores. Langmuir **1998**, 14, 4435–4444.

83. Di Carlo, D.; Irimia, D.; Tompkins, R.G.; Toner, M. Continuous inertial focusing, ordering, and separation of particles in microchannel. Proc. Natl. Acad. Sci. USA **2007**, 104, 18892–18897.

84. Gunther, A.; Jensen, K.F. Multiphase microfluidics: From flow characteristics to chemical and materials synthesis. Lab Chip **2006**, 6, 1487–1503.

85. Gunther, A.; Jhunjhunwala, M.; Thalmann, M.; Schmidt, M.A.; Jensen, K.F. Micromixing of miscible liquids in segmented gas-liquid flow. Langmuir **2005**, 21, 1547–1555.

86. Gunther, A.; Khan, S.A.; Thalmann, M.; Trachsel, F.; Jensen, K.F. Transport and reaction in microscale segmented gas-liquid flow. Lab Chip **2004**, 4, 278–286.

87. Wang, F.K.; Chon, C.H.; Li, D.Q. Particle separation by a moving air-liquid interface in a microchannel. J. Colloid Interface Sci. **2010**, 352, 580–584.

88. Subramaniam, A.B.; Abkarian, M.; Mahadevan, L.; Stone, H.A. Mechanics of interfacial composite materials. Langmuir **2006**, 22, 10204–10208.

89. Subramanian, R.S.; Larsen, R.J.; Stone, H.A. Stability of a flat gas-liquid interface containing nonidentical spheres to gas transport: Toward an explanation of particle stabilization of gas bubbles. Langmuir **2005**, 21, 4526–4531.

90. Subramaniam, A.B.; Abkarian, M.; Stone, H.A. Controlled assembly of jammed colloidal shells on fluid droplets. Nat. Mater. **2005**, 4, 553–556.

91. Subramaniam, A.B.; Mejean, C.; Abkarian, M.; Stone, H.A. Microstructure, morphology, and lifetime of armored bubbles exposed to surfactants. Langmuir **2006**, 22, 5986–5990.

92. Lautz, J.; Sankin, G.; Yuan, F.; Zhong, P. Displacement of particles in microfluidics by laser-generated tandem bubbles. Appl. Phys. Lett. **2010**, 97, 183701–183701-3.

93. Wang, M.M.; Tu, E.; Raymond, D.E.; Yang, J.M.; Zhang, H.C.; Hagen, N.; Dees, B.; Mercer, E.M.; Forster, A.H.; Kariv, I.; et al. Microfluidic sorting of mammalian cells by optical force switching. Nat. Biotechnol. **2005**, 23, 83–87.

94. Durr, M.; Kentsch, J.; Muller, T.; Schnelle, T.; Stelzle, M. Microdevices for manipulation and accumulation of micro- and nanoparticles by dielectrophoresis. Electrophoresis **2003**, 24, 722–731.

95. Gijs, M.A.M. Magnetic bead handling on-chip: New opportunities for analytical applications. Microfluid. Nanofluid.**2004**, 1, 22–40.

96. Lu, L.H.; Ryu, K.S.; Liu, C. A magnetic microstirrer and array for microfluidic mixing. J. Microelectromech. Syst. **2002**,11, 462–469.

97. Challis, R.E.; Povey, M.J.W.; Mather, M.L.; Holmes, A.K. Ultrasound techniques for characterizing colloidal dispersions. Rep. Prog. Phys. **2005**, 68, 1541–1637.

98. Lilliehorn, T.; Simu, U.; Nilsson, M.; Almqvist, M.; Stepinski, T.; Laurell, T.; Nilsson, J.; Johansson, S. Trapping of microparticles in the near field of an ultrasonic transducer. Ultrasonics **2005**, 43, 293–303.

99. Mason, W.P. Physical Acoustics; Academic Press: New York, NY, USA, 1982.

100. Poesio, P.; Ooms, G. Formation and ultrasonic removal of fouling particle structures in a natural porous material. J. Pet. Sci. Eng. **2004**, 45, 159–178.

101. Spengler, J.; Jekel, M. Ultrasound conditioning of suspensions—Studies of streaming influence on particle aggregation on a lab- and pilot-plant scale. Ultrasonics **2000**, 38, 624–628.

102. Laurell, T.; Petersson, F.; Nilsson, A. Chip integrated strategies for acoustic separation and manipulation of cells and particles. Chem. Soc. Rev. **2007**, 36, 492–506.

103. Sedelmeier, J.; Ley, S.V.; Baxendale, I.R.; Baumann, M. KMnO(4)-mediated oxidation as a continuous flow process.Org. Lett. **2010**, 12, 3618–3621.

104. Browne, D.L.; Deadman, B.J.; Ashe, R.; Baxendale, I.R.; Ley, S.V. Continuous flow processing of slurries: Evaluation of an agitated cell reactor. Org. Proc. Res. Dev. **2011**, 15, 693–697.

Chapter 7

NANO-CATALYSTS WITH MAGNETIC CORE: SUSTAINABLE OPTIONS FOR GREENER SYNTHESIS

Rajender S Varma

Sustainable Technology Division, National Risk Management Research Laboratory, U.S. Environmental Protection Agency, 26 West Martin Luther King Drive, MS 443, Cincinnati, Ohio 45268, USA

ABSTRACT

Author's perspective on nano-catalysts with magnetic core is summarized with recent work from his laboratory. Magnetically recyclable nano-catalysts and their use in benign media is an ideal blend for the development of sustainable methodologies in organic synthesis. Water or polyethylene glycol (PEG) provides good medium to perform such chemical reactions with magnetic nano-catalysts, as this combination adds exceptional value to the overall sustainable process development. In this mini-review, the uses of magnetically recyclable nano-catalysts for a variety of organic reactions are described in conjunction with activation via microwave irradiation.

Chemical developments in the new millennium are now routinely utilizing the concept of "green chemistry" to meet the challenges of protecting the environment and human health while maintaining commercial viability. *Green Chemistry* is defined as "the utilization of a set of principles that eliminates or reduces the use or generation of hazardous substances in the design, manufacture, and application of chemical products" and emphasizes hazard reduction as the performance criteria while designing new chemical processes. Nanotechnology processes in the recent years have enabled paradigm changing developments in environmental science, medicine and importantly, catalysis. One of the key areas for achieving sustainability is to explore the generation of efficient catalytic processes, via nano-catalysis [1] as a viable option; catalysis has been an integral part of what is now defined more clearly under nanotechnology domain [2].

Nanoparticles-Greener Synthesis

Nanoparticles are the miniscule building blocks for an array of new commercial products and consumer materials in the emerging field of nanotechnology; they are discovered and introduced in the market place at a very fast pace. Nanoparticles can be defined as particulate matter with at least one dimension that is less than 100 nm. The commercial interest in nanotechnology has significantly increased with more than US $9 billion in investment from public and private sources [3]. Nanoparticles have an exceptionally large surface area to volume ratio, an important trait that is responsible for their widespread advantageous use in catalysis. The nanoscale size and shape imparts unique properties to catalysts because of the structural and electronic changes which differentiates them from the bulk materials. The fine tuning of nano-catalysts, in terms of composition (bimetallic, core-shell type or use of supports), shape and size has accomplished greater selectivity.

In the catalysis arena, the production of engineered nanomaterials is a major breakthrough in material science. A sustained effort has been made to develop eco-friendly strategies to generate these nanomaterials via pathways that use benign reagents rather than the hazardous substances conventionally used. The sustainable strategy for the preparation of nanoparticles has been exemplified by the use of vitamins B_1, B_2, C, and tea and wine polyphenols [4], which function both as reducing and capping agents [5]. This obviates the need to use toxic reducing agents, such as borohydrides or hydrazines. These extremely simple and aqueous green synthetic methods generate bulk quantities of nano-catalysts without the need for large amounts of insoluble templates [6] and have found numerous applications in catalysis [1, 5, 7].

Alternate energy input- Use of microwave irradiation

The desired approach to the preparation of uniformly small-sized nano-catalysts may include alternative activation methodology, such as microwave (MW)-, and ultrasonic irradiation and mechanochemical mixing [8]. The rapid and in-core MW heating has proven a useful method for the synthesis of these metallic nanostructures in solutions [9, 10]. The valuable application of this method has been established for the preparation of silver (Ag), gold (Au), platinum (Pt), and gold-palladium (Au-Pd) nanostructures [5, 11]. MW heating conditions not only allow the rapid preparation of spherical nanoparticles, but also enables the formation of single crystalline polygonal plates, sheets, rods, wires, tubes, and dendrites as well [12]. Nanostructures of uniformly smaller sizes and with narrower size distributions and a higher degree of crystallization have been obtained using MW heating than those prepared via the conventional oil-bath heating; MW approach allows the greener synthesis of nanomaterials with several desirable features, such as shorter reaction times, better product yields and reduced energy consumption (Figure 1) [7].

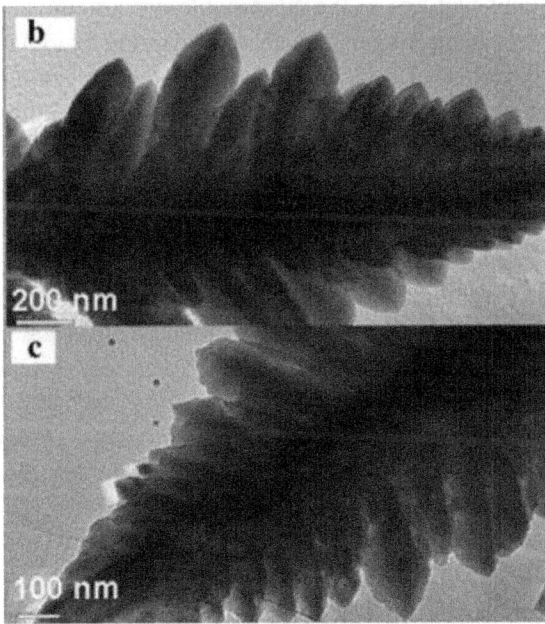

Figure 1: MW Synthesis of dendritic nano- ferrites (micro- pine morphology) from inexpensive starting materials in water without using any reducing or capping reagent agent; panels b and c show the well-defined and highly ordered branches. (Reproduced by permission from Royal Society of Chemistry, reference 7).

Importantly, the strategy encompasses *"benign by design"* principles and focus on the utilization of renewable resources, if possible [4, 5].

Nanoparticles with Magnetic Core

To protect the environment and to avoid undue manipulation efforts to purify the product and separation of the catalysts, there is a pressing need to develop methodologies that can facilitate recycle and reuse of these materials at low concentrations and in complex matrices. In this context, iron-based magnetic nanoparticles (MNPs) have been developed that address these needs [13]. These MNPs can be further divided based on the nature of the magnetic core, which can be made up of either reduced species or oxides such as iron oxide NPs (Fe_2O_3 and Fe_3O_4) that has found application in oxidative and coupling reactions [14, 15]. Such unmodified bare iron oxides comprising Fe_3O_4 and Fe_2O_3 are found to be active catalysts for the coupling of aldehyde, alkyne, and amine (A^3 coupling) to afford an easy route to propargylamines [16].

The use of MNPs as catalysts in chemical synthesis has been extensively studied in recent years as the recovery of expensive catalysts after their use are some of the salient features in the sustainable process development [17–21]. MNPs coated with benign ligands such as dopamine or glutathione have also been developed and used as heterogeneous catalysts for numerous organic transformations and syntheses [17–21]. The functionalization of the surfaces of nano-sized magnetic materials provides a quasi-homogeneous phase in reaction media and acts as a bridge between heterogeneous and homogeneous catalysis thus retaining the relative virtues of both of the systems.

The greener generation of nanoparticles and their eco-friendly applications in catalysis via magnetically recoverable and recyclable nano-catalysts for a variety of oxidation, reduction, and condensation reactions [20–25], has made a tremendous impact on the development of sustainable pathways. This heterogenization of the catalyst in the form of MNPs allows them to be recovered using an external magnet and facilitate their subsequent reuse more efficiently. In view of the reduced size of MNPs in nm range, most of the catalysts surface is accessible for reaction as it provides quasi homogeneous media for the catalysts.

Synthetic Modification of Magnetic Nanoparticles

The synthetic modification of nanoferrites with dopamine and subsequent anchoring of the metal particles (Scheme 1) on its surface provides numerous opportunities to deploy these nano-catalysts effectively. The hydration of benzonitrile with ruthenium hydroxide on magnetic nano-ferrites transforms it to benzamide in water is an example [22]. The catalyst can be easily separated using an external magnet and after its separation, the clear reaction mixture could be cooled to generate the crystals of benzamides. The complete procedure could be conducted exclusively in aqueous medium that is devoid of organic solvents [22]; the MW-assisted reaction proceeds with high turnover numbers and is attributed to the use of a nano-catalyst.

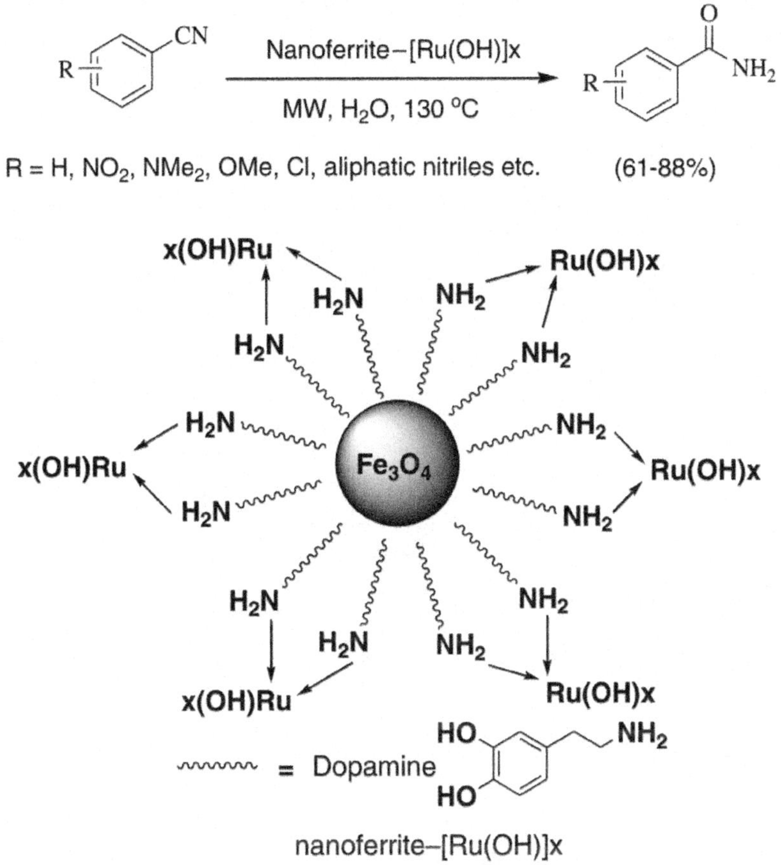

Scheme 1: Nanoferrite–[Ru(OH)]x catalyzed aqueous hydration of nitriles (reproduced by permission from Royal Society of Chemistry, reference 7).

Nanoferrite–[Ru(OH)]x Catalyzed Aqueous Hydration of Nitriles (reproduced by permission from Royal Society of Chemistry, reference 7).

In an analogous manner, ubiquitous glutathione (GT) molecules have been immobilized on magnetic nanoferrites via their thiol groups [20] and the resulting catalyst (nano-FGT) has been used for the efficient synthesis of a wide variety of aryl, alkyl, and heterocyclic amines (Scheme 2). Such organocatalytic approach enables facile conversion of functionalized amines selectively into the corresponding pyrroles without affecting several sensitive functional groups [25]. This salient feature of immobilization of organocatalysts on MNPs enables their separation magnetically, after completion of the reaction, thus avoiding the cumbersome traditional separation by chromatographic means and solvent usage.

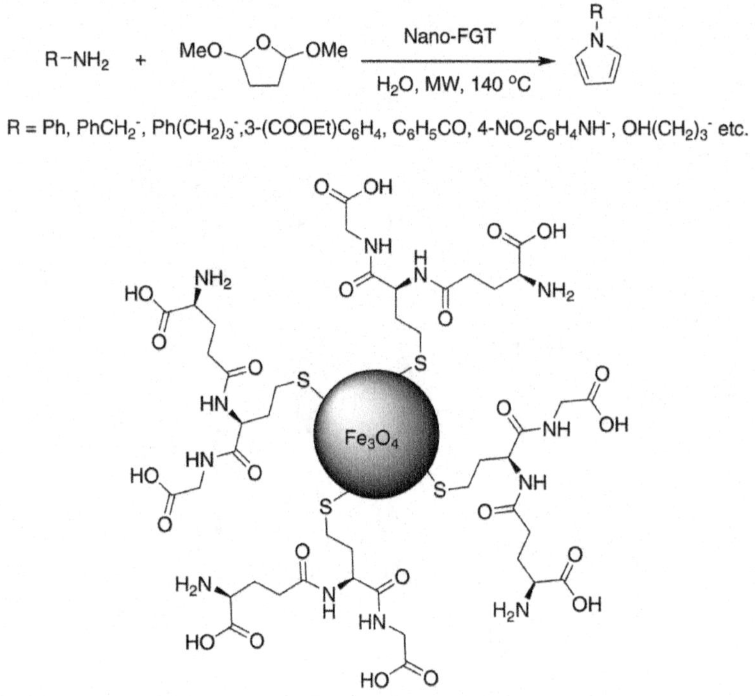
R−NH2 +
MeO
OMe
Nano-FGT
H2O, MW, 140 °C
R = Ph, PhCH2-, Ph(CH2)3-,3-(COOEt)C6H4, C6H5CO, 4-NO2C6H4NH-, OH(CH2)3- etc.
Fe3O4

Scheme 2: Nano-FGT-catalyzed Paal–Knorr reactions (reproduced by permission from Royal Society of Chemistry, reference 7)

Nano-FGT-catalyzed Paal–Knorr reactions (reproduced by permission from Royal Society of Chemistry, reference 7).

Such nanoparticle-supported and magnetically recoverable organocatalysts help catalyze the synthesis of heterocyclic entities (Paal-Knorr and other reactions) in pure aqueous medium thus precluding the use of toxic organic solvents, even in the work-up stages [1, 7, 25].

Magnetically separable nano-FGT-Cu catalyst could be used for azide alkyne cycloaddition (AAC) reaction and this general reaction is performed in one-pot via *in situ* azide generation followed by cycloaddition in aqueous media (Scheme 3) [26].

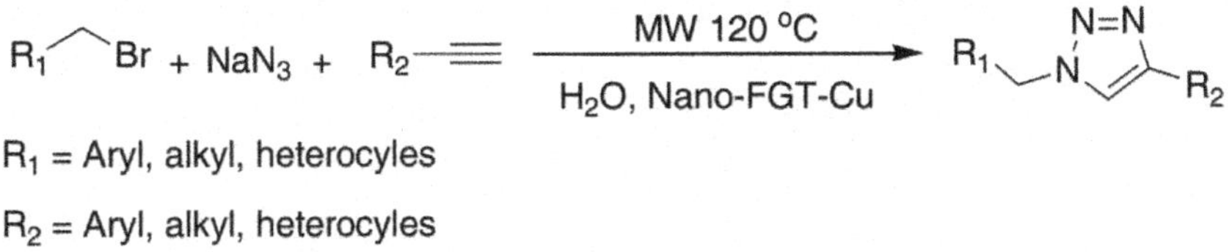

$$R_1 \diagup Br + NaN_3 + R_2 - \!\!\!\equiv \quad \xrightarrow[\text{H}_2\text{O, Nano-FGT-Cu}]{\text{MW 120 °C}} \quad R_1 \diagdown N \diagup^{N=N}_{} \diagdown R_2$$

R_1 = Aryl, alkyl, heterocyles

R_2 = Aryl, alkyl, heterocyles

Scheme 3: Nano-FGT-Cu catalyzed 1,3-dipolar cycloadditions reaction (reproduced by permission from Royal Society of Chemistry, reference 7)

Nano-FGT-Cu catalyzed 1,3-dipolar cycloadditions reaction (reproduced by permission from Royal Society of Chemistry, reference 7).

A similar nano-FeDOPACu bimetallic catalyst (Figure 2) has been deployed in the C-S coupling of aryl halides with thiophenols under MW irradiation conditions [27]; several iodides and bromides (with the exception of aryl chlorides) undergo reaction with thiophenols, affording the corresponding diaryl sulfides (Scheme 4).

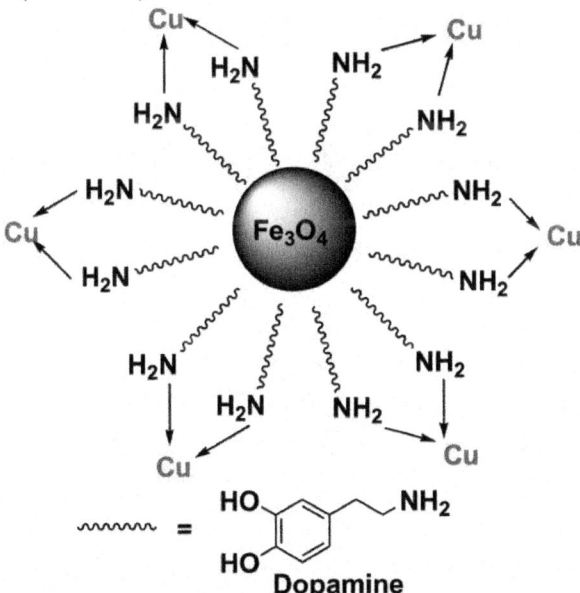

Figure 2: Nano- Fe_3O_4 -DOPA-Cu catalyst (nano- FeDOPACu) (reproduced by permission from Royal Society of Chemistry, reference 7).

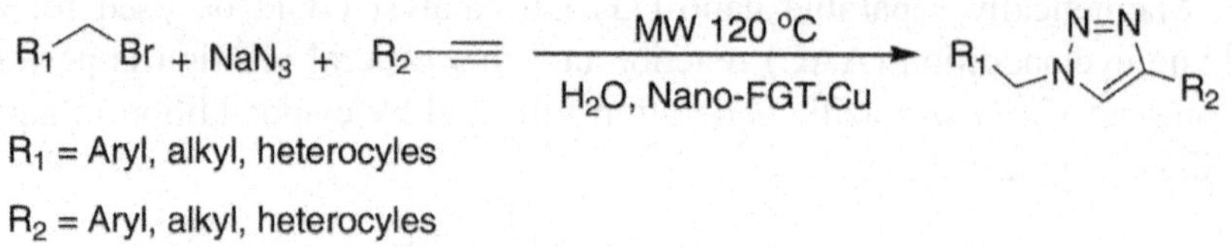

R_1 = Aryl, alkyl, heterocyles

R_2 = Aryl, alkyl, heterocyles

Scheme 4: Nano-FeDOPACu catalyzed cross coupling of aryl halides with thiophenols

Nano-FeDOPACu Catalyzed Cross Coupling of Aryl Halides with Thiophenols.

The activity of a Fe-Cu bimetallic catalyst can, however, be altered by changing the anchoring ligand for the immobilization of copper nanoparticles on magnetic nano ferrite (Fe_3O_4) surface. When glutathione is used, the catalyst shows activity for Huisgen cycloaddition [26]. In contrast, when dopamine is used as the anchoring ligand for Cu nanoparticles, the ensuing catalyst is active for C-S coupling and completely inactive for the Huisgen cycloaddition reactions [28].

A conceptually simple strategy has been advanced for synthetic chemists wherein synthesis of $Fe_3O_4@SiO_2Ru$ for the hydration of nitriles in aqueous media occurs (Scheme 5) via sequential addition of reagents in one-pot to generate and to coat the magnetic particles [29].

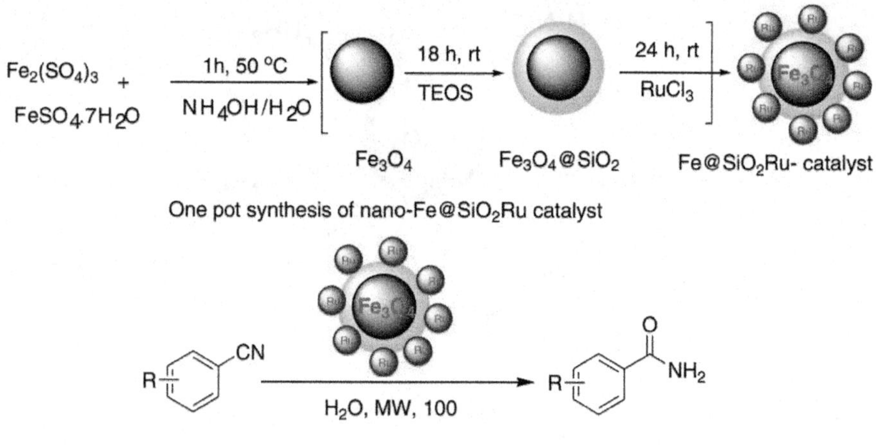

One pot synthesis of nano-Fe@SiO2Ru catalyst

Scheme 5: Hydration of nitriles by nano-Fe@SiO$_2$Ru catalyst (reproduced by permission from Royal Society of Chemistry, reference 7)

Hydration of Nitriles by Nano-Fe@SiO$_2$Ru Catalyst (reproduced by permission from Royal Society of Chemistry, reference 7).

Simple mesoporous silica supported iron oxide nanoparticles have been used for the oxidation of alkenes using hydrogen peroxide in water [30]. Styrenes and their substituted derivatives are converted to respective aldehydes in excellent yield (Scheme 6). It is important to note that no other side products or oxidation of aldehydes are observed during the reaction. The optimized protocol is suitable to a range of electron-withdrawing and electron-donating substituents localized in all positions of the aromatic ring.

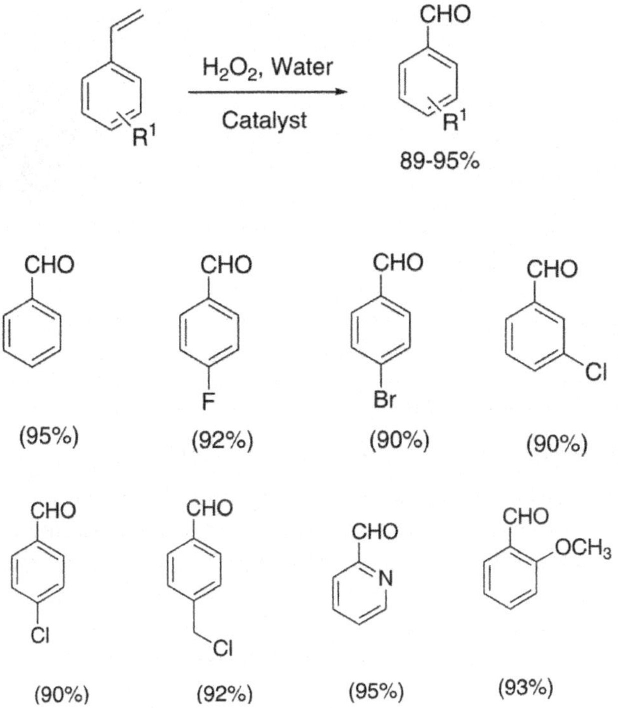

Scheme 6: Mesoporous silica supported iron oxide nanoparticles for the oxidation of alkenes using hydrogen peroxide in aqueous medium

Mesoporous Silica Supported Iron Oxide Nanoparticles for the Oxidation of Alkenes Using Hydrogen Peroxide in Aqueous Medium

Supported iron oxide nanocatalysts have been the focus in various important catalytic applications because of their low cost and toxicity, ready availability, and environmentally benign nature [7, 31].

Asymmetric synthesis using magnetic nanoparticles

The utility of MNP supported catalysts in asymmetric synthesis has been illustrated in an example where superparamagnetic nanoparticle-supported (S)-diphenylprolinol trimethylsilyl ether has been employed for the asymmetric Michael addition in water [32]; Jørgensen–Hayashi catalyst, ((S)-diphenylprolinol trimethylsilyl ether) on silica superparamagnetic nanoparticle support, (Figure 3), was prepared by a multistep synthetic procedure.

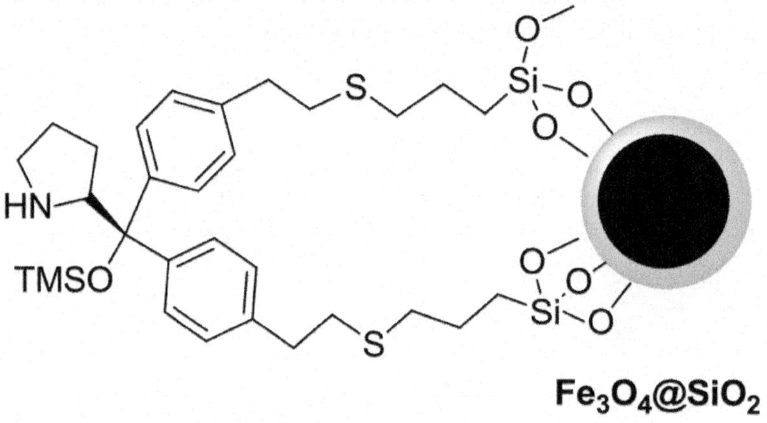

$Fe_3O_4@SiO_2$

Figure 3: MNP- supported Jørgensen– Hayashi Catalyst.

As-synthesized MNP-supported Jørgensen–Hayashi catalyst accomplished the asymmetric Michael addition of enolisable aldehydes to nitroalkenes in water; the corresponding products were obtained in moderate to good yields (up to 96%), and with good enantioselectivity (up to 90% ee) (Scheme 7).

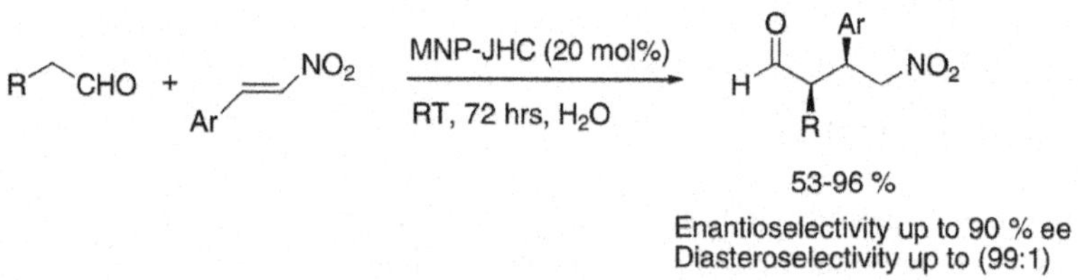

53-96 %
Enantioselectivity up to 90 % ee
Diasteroselectivity up to (99:1)

Scheme 7: Aqueous asymmetric Michael addition of aldehydes to nitroalkenes.

Aqueous Asymmetric Michael Addition of Aldehydes to Nitroalkenes

Significantly, the reaction was conducted between trans- -nitrostyrene and propanal in water to deliver high yield of the product (85%). A variety of nitroalkanes and aldehydes performed well under these optimized reactions conditions and the corresponding products were obtained in good to excellent yields in benign aqueous medium.

Synthetic processes using alternative energy input in combination with nano-catalysts thus shorten the reaction time and that eliminate or minimize side product formation [8]. This concept when used in the syntheses of pharmaceuticals, fine chemicals, and polymers may pave the way towards the greener and more sustainable approach to chemical syntheses. Newer developments on these themes, especially involving benign reaction media such as water and polyethylene glycol (PEG), in conjunction with photo activation [33], MW and ultrasonic irradiation, and/or ball-milling under solvent-free conditions, may help realize sustainable pathways for chemical synthesis and transformations, including the generation of novel nano-catalysts [7, 8].

Nano-catalysts, especially those that can be recycled magnetically, are of special value in synthetic domain. The modifications of the surface of MNPs with other nanometals are of interest for catalysis. Such post-synthetic alterations using organic ligands [20], mimicking organocatalysis, allows the adsorption of catalytically active metal nanoparticles, which provide identical or better reactivity than the corresponding homogeneous catalysts [25, 34]. The enhanced dispersity of the MNPs in common solvents is an added advantage, since it exposes the surface-bound active reaction sites for the reactants in ideal fashion. In foreseeable future, the design of novel magnetically retrievable heterogeneous asymmetric catalysts supported on Fe_3O_4 nanoparticle systems will find significant applications in asymmetric hydrogenations [35], asymmetric C-C bond formation reactions, and in asymmetric cycloaddition reactions, especially under continuous-flow conditions in microreactors [36].

Bare iron oxide nanoparticles provide ready access to oxidation and oxidative coupling reactions, while the reduced Fe(0) NPs facilitate dehydrogenation, hydrogenation, couplings and reductive processes. The incorporation of a second metal serves to futher expand the catalytic abilities of Fe and incorporation of oxides on to silica has been utilized for the biodiesel production [37]. Supporting various ligands on the magnetic nanoparticles facilitates synthesis of several chemical entities with sustainable advantages, predominant being the magnetic separation and recyclability aspects [38, 39]. Incorporation of copper provides an easy preparation of propargylamines via a multicomponent reaction [40] and asymmetric nanocatalysis is facilitated by support of heterocyclic carbenes as chiral modifiers [41].

Magnetic nanoparticles are garnering special attention in the emerging area of flow chemistry. The dual function of confinement and agitation of the nanoparticle-bound catalyst can be achieved in a reactor by means of a rotating magnetic field, which avoids the potential problems of clogging membranes or filters that are eminent barriers for immobilized catalysts. Reiser et al. have demonstrated this successfully in a close circuit reactor for the asymmetric benzoylation of racemic 1,2-diols using a copper(II)-azabis(oxazoline) catalyst, which had been covalently attached to carbon-coated cobalt nanoparticles [42]; magnetic field induced flow mixing is beginning to make critical impact for handling of slurries and precipitates in modern small footprint flow reactors [43].

ACKNOWLEDGEMENTS

I wish to thank my collaborators and colleagues, past and present, whose names appear in the reference section for their immense contribution to our efforts in Green and Sustainable Chemistry research program.

REFERENCES

1.　Polshettiwar V, Varma RS: Green chemistry by nano-catalysis. *Green Chem* 2010, 12:743–754.

2.　Astruc D, Lu F, Aranzaes JR: Nanoparticles as recyclable catalysts: The frontier between homogeneous and heterogeneous catalysis. *Angew Chem Inter Edn:* 2005, 44:7852–7872.

3.　Eckelman MJ, Zimmerman JB, Anastas PT: Toward green nano: E-factor analysis of several nanomaterial syntheses. *J Ind Ecol*2008, 12:316–328.

4.　Baruwati B, Varma RS: High value products from waste: grape pomace extract - a three-in-one package for the synthesis of metal nanoparticles. *ChemSusChem* 2009, 2:1041–1044.

5.　Varma RS: Greener approach to nanomaterials and their sustainable applications. *Curr Opin Chem Eng* 2012, 1:123–128.

6.　Nadagouda MN, Speth T, Varma RS: Microwave-assisted green synthesis of silver nanostructures. *Acc Chem Res* 2011, 44:469–478.

7.　Varma RS: Journey on greener pathways: From the use of alternate energy inputs and benign reaction media to sustainable applications of nano-catalysts in synthesis and environmental remediation. *Green Chem* 2014, 16:2027–2047.

8.　Nasir Baig RB, Varma RS: Alternate energy input: mechanochemical, microwave and ultrasound-assisted organic synthesis.*Chem Soc Rev* 2012, 41:1559–1584.

9. Polshettiwar V, Nadagouda MN, Varma RS: Microwave-assisted chemistry: A rapid and sustainable route to synthesis of organics and nanomaterials. *Aust J Chem* 2009, 62:16–26.

10. Baghbanzadeh M, Skapin SD, Orel ZC, Kappe CO: Critical assessment of the specific role of microwave irradiation in the synthesis of ZnO micro- and nano-structure materials. *Chem Eur J* 2012, 18:5724–5731.

11. Varma RS: Green chemical processes: sustainable nanomaterials from waste agricultural residues and their applications.*Speciality Chemicals Magazine* 2013, 33:27–28.

12. Polshettiwar V, Baruwati B, Varma RS: Self-assembly of metal oxides into three-dimensional nanostructures: synthesis and application in catalysis. *ACS Nano* 2009, 3:728–736.

13. Baruwati B, Nadagouda MN, Varma RS: Bulk synthesis of monodisperse ferrite nanoparticles at water-organic interfaces under conventional and microwave hydrothermal treatment and their surface functionalization. *J Phys Chem C* 2008, 112:18399–18404.

14. Enthaler S, Junge K, Beller M: Sustainable metal catalysis with iron: from rust to a rising star? *Angew Chem Int Edn* 2008, 47:3317–3321.

15. Bolm C, Legros J, Le Paih J, Zani L: Iron-catalyzed reactions in organic synthesis. *Chem Rev* 2004, 104:6217–6254.

16. Zeng TQ, Chen WW, Cirtiu CM, Moores A, Song GH, Li CJ: Fe_3O_4 nanoparticles: a robust and magnetically recoverable catalyst for three-component coupling of aldehyde, alkyne and amine. *Green Chem* 2010, 12:570–573.

17. Baig RBN, Varma RS: Magnetically retrievable catalysts for organic synthesis. *Chem Commun* 2013, 49:752–770.

18. Gawande MB, Rathi AK, Branco PS, Varma RS: Sustainable utility of magnetically recyclable nano-catalysts in water: applications in organic synthesis. *Appl Sci* 2013, 3:656–674.

19. Gawande MB, Branco PS, Varma RS: Nano-magnetite (Fe_3O_4) as a support for recyclable catalysts in the development of sustainable methodologies. *Chem Soc Rev* 2013, 42:3371–3393.

20. Polshettiwar V, Baruwati B, Varma RS: Magnetic nanoparticle-supported glutathione: A conceptually sustainable organocatalyst.*Chem Commun* 2009, 1837–1839.

21. Baig RBN, Varma RS: Organic synthesis via magnetic attraction: benign and sustainable protocols using magnetic nanoferrites.*Green Chem* 2013, 15:398–417.

22. Polshettiwar V, Varma RS: Nanoparticle-supported and magnetically recoverable ruthenium hydroxide catalyst: Efficient hydration of nitriles to amides in aqueous medium. *Chem Eur J* 2009, 15:1582–1586.

23. Polshettiwar V, Baruwati B, Varma RS: Nanoparticle-supported and magnetically recoverable nickel catalyst: a robust and economic hydrogenation and transfer hydrogenation protocol. *Green Chem* 2009, 11:127–131.

24. Luque R, Baruwati B, Varma RS: Magnetically separable nanoferrite-anchored glutathione: aqueous homocoupling of arylboronic acids under microwave irradiation. *Green Chem* 2010, 12:1540–1543.

25. Polshettiwar V, Varma RS: Nano-organocatalyst: magnetically retrievable ferrite-anchored glutathione for microwave-assisted Paal-Knorr reaction, aza-Michael addition, and pyrazole synthesis. *Tetrahedron* 2010, 66:1091–1097.

26. Baig RBN, Varma RS: A highly active magnetically recoverable nano ferrite-glutathione-copper (nano-FGT-Cu) catalyst for Huisgen 1,3-dipolar cycloadditions. *Green Chem* 2012, 14:625–632.

27. Baig RBN, Varma RS: A highly active and magnetically retrievable nanoferrite-DOPA-copper catalyst for the coupling of thiophenols with aryl halides. *Chem Commun* 2012, 48:2582–2584.

28. Baig RBN, Varma RS: Copper modified magnetic bimetallic nano-catalysts: ligand regulated catalytic activity. *Curr Org Chem* 2013, 17:2227–2237.

29. Baig RBN, Varma RS: A facile one-pot synthesis of ruthenium hydroxide nanoparticles on magnetic silica: aqueous hydration of nitriles to amides. *Chem Commun* 2012, 48:6220–6222.

30. Rajabi F, Karimi N, Saidi MR, Primo A, Varma RS, Luque R: Unprecedented selective oxidation of styrene derivatives using a supported iron oxide nanocatalyst in aqueous medium. *Adv Synth Catal* 2012, 354:1707–1711.

31. Polshettiwar V, Baruwati B, Varma RS: Nanoparticle-supported and magnetically recoverable nickel catalyst: a robust and economic hydrogenation and transfer hydrogenation protocol. *Green Chem* 2009, 11:127–131.

32. Wang BG, Ma BC, Wang Q, Wang W: Superparamagnetic nanoparticle-supported (s)-diphenylprolinol trimethylsilyl ether as a recyclable catalyst for asymmetric Michael addition in water. *Adv Synth Catal* 2010, 352:2923–2928.

33. Balu AM, Baruwati B, Serrano E, Cot J, Garcia-Martinez J, Varma RS, Luque R: Magnetically separable nanocomposites with photocatalytic activity under visible light for the selective transformation of biomass-derived platform molecules. *Green Chem*2011, 13:2750–2758.

34. Varma RS: Greener routes to organics and nanomaterials: sustainable applications of nano-catalysts. *Pure & Applied Chem* 2013,85:1703–1710.

35. Hu A, Yee GT, Lin W: Magnetically recoverable chiral catalysts immobilized on magnetite nanoparticles for asymmetric hydrogenation of aromatic ketones. *J Am Chem Soc* 2005, 127:12486–12487.

36. Vaddula BR, Varma RS, Gonzalez MA: Supported ruthenium catalysts for sustainable flow chemistry. *Curr Org Chem* 2013,17:2268–2278.

37. Suzuta T, Toba M, Abe Y, Yoshimura Y: Iron oxide catalysts supported on porous silica for the production of biodiesel from crude Jatropha oil. *J Amer Oil Chem Soc* 2012, 89:1981–1989.

38. O› Dalaigh C, Corr SA, Gun›ko Y, Connon SJ: A magnetic-nanoparticle-supported 4-N,N-dialkylaminopyridine catalyst: Excellent reactivity combined with facile catalyst recovery and recyclability . *Angew Chem Inter Edn* 2007, 46:4329–4332.

39. Li PH, Li BL, An ZM, Mo LP, Cui ZS, Zhang ZH: Magnetic Nanoparticles (CoFe$_2$O$_4$)-supported phosphomolybdate as an efficient, green, recyclable catalyst for synthesis of β-hydroxy hydroperoxides. *Adv Synth Catal* 2013, 355:2952–2959.

40. Aliaga MJ, Ramon DJ, Yus M: Impregnated copper on magnetite: an efficient and green catalyst for the multicomponent preparation of propargylamines under solvent free conditions. *Org & Biomol Chem* 2010, 8:43–46.

41. Ranganath KVS, Kloesges J, Schafer AH, Glorius F: Asymmetric nanocatalysis: N-heterocyclic carbenes as chiral modifiers of Fe$_3$O$_4$/Pd nanoparticles. *Angew Chem Inter Edn* 2010, 49:7786–7789.

42. SchatzA,GrassRN,KainzQ,StarkWJ,ReiserO:Cu(II)-azabis(oxazoline) complexes immobilized on magnetic Co/C nanoparticles: Kinetic resolution of 1,2-diphenylethane-1,2-diol under batch and continuous-flow conditions. *Chem Mater* 2010,22:305–310.

43. Koos P, Browne DL, Ley SV: Continuous stream processing: a prototype magnetic field induced flow mixer. *Green Process Synth*2012, 1:11–18.

Chapter 8

DEVELOPMENT OF A SUSTAINABLE PROCESS FOR THE PRODUCTION OF POLYMER GRADE LACTIC ACID

Susmit S Bapat, Clint P Aichele and Karen A High

School of Chemical Engineering, Oklahoma State University, Stillwater, OK 74078, USA

ABSTRACT

Lactic acid is a commonly occurring substance in nature, ranging from existence in micro-organisms to the human body. Traditionally, lactic acid has applications in industries such as food, chemicals, pharmaceuticals and textiles. In this work, a sustainable process for the production of polymer grade lactic acid (99 wt. % on dry basis) from crude lactic acid was simulated. The simulation was performed using Aspen Plus® version 8.2. The thermodynamic model used for the process was NRTL – Hayden O'Connell due to the polar nature and non-ideal behavior of the species involved. The process was carried out in three stages. First, crude lactic acid was obtained by reacting calcium lactate with sulfuric acid. The second stage consisted of esterification of lactic acid by reactive distillation. A RadFrac column was used for this purpose which also facilitated easy separation of methyl lactate from methanol and water. Pure methyl lactate obtained from the second stage was then hydrolyzed in the third stage using pure lactic acid as an auto-catalyst to obtain the desired product. Use of pure lactic acid as an auto-catalyst helped to achieve the required purity as it minimized contamination. The process was optimized using sensitivity analysis in Aspen Plus®.

INTRODUCTION

Lactic acid was first isolated in 1780 by Swedish scientist Carl Wilhelm Scheele by crystallizing its calcium salt [1]. It is a weak organic acid with a hydroxyl group and a carboxylic group present on adjacent carbon atoms in the carbon chain. This duality in structure allows it to react either as an acid or as an alcohol.

Traditionally, lactic acid has applications in the food, chemical, and pharmaceutical industries. In the food industry, lactic acid with a purity of about 85 wt. % is used to induce a sour taste in food products, for example in pickles and sauerkraut. It is also used as an acidulant in the food industry [2]. In addition, it is also used in the textile industry as a caustic [3, 4]. Around 80–85 wt. % of lactic acid is manufactured for use in the food industry. More recently however, there has been an increased focus on manufacturing lactic acid of high purity (99 wt. % on dry basis) which can be used as the monomer for producing poly-lactic acid (PLA). PLA can be obtained from lactic acid via two different mechanisms – 1) direct polymerization of lactic acid by poly-condensation or 2) condensation to form lactide (intermediate) and corresponding ring-opening polymerization to obtain PLA [3].

PLA is a biocompatible and biodegradable material that has numerous applications in sustainable plastic products [5]. Along with ease of disposal, lactic acid polymers also possess high tensile strength and can be used in the packaging industry and for medical and biological applications [6]. Table 1 summarizes the applications of PLA in various industries [2, 7].

Table 1: Summary of polymer grade lactic acid applications

Industry	Property	Application
Medical	Non-toxic, relatively strong, bio-compatible, sterilizable	Medical Implants; clinical applications – sutures; meshes, bone fixation devices
Packaging	High tensile strength, thermal resistance, impact resistance, transparency	Flexible Films, thermoforming, lamination
Textile	bacteriostatic, flame-retardant, and weathering resistance	Geo-textiles, Industrial Fabrics, Fibers, Home Furnishings
Environmental	Bio-compatible	As sorbent in wastewater treatment; As a substrate for nitrogen removal; as a bioremediation agent
Other	Biodegradability, flame-retardant, thermal resistance	To manufacture sandbags, weed prevention nets, vegetation nets, vegetation pots, ropes, binding tape for use in the agriculture industry

PLA competes with traditional plastics like poly-ethylene terephthalate (PET) and poly propylene (PP) in terms of sustainability [8] and applications. PLA is a biodegradable polymer and decomposes at composting conditions and temperatures above 60°C. PLA production requires less energy per kg as compared to PET and PP (42 MJ/kg for PLA as compared to 73 MJ/kg for PP and 80 MJ/kg for PET) [8]. Recycling PLA, however, is difficult using traditional mechanical or melt-recycling methods because of its temperature and water sensitivity. To accomplish this, chemical processes which can hydrolyze PLA to lactic acid are being developed.

According to a comparison study performed by Tabone, Cregg et. al., PLA is the top ranked polymer in terms of rankings based on green design principles and ranks sixth according to Life Cycle Assessment principles [9]. It is therefore considered as one of the more sustainable alternatives to plastics being used currently.

To obtain high quality PLA, polymer grade lactic acid (~99 wt. % purity) is the starting point. Several efforts have been made previously to obtain high purity lactic acid [10–12]. Those methods generally have the following limitations.

- High pressure and temperature required to achieve the intended purity which increases cost.
- High pressure and temperature also lead to formation of by-products (unwanted methyl esters) during the esterification reaction which are difficult to separate.
- Use of acid catalyst in the hydrolysis reaction hampers lactic acid purity.
- Limited conversion for the esterification reaction due to product build-up.

Process Description

The present simulation work is based on a patented process by the National Chemical Laboratory (Pune) to manufacture polymer grade lactic acid [13]. The process can be roughly divided into three stages. The first stage consists of reacting 10 wt. % calcium lactate (obtained as a fermentation product) with concentrated H_2SO_4 to yield dilute crude lactic acid and calcium sulfate as shown in Figure 1. The sulfate is separated out by means of a centrifuge and the crude lactic acid is concentrated to 60 wt. % by passing it through a falling film evaporator.

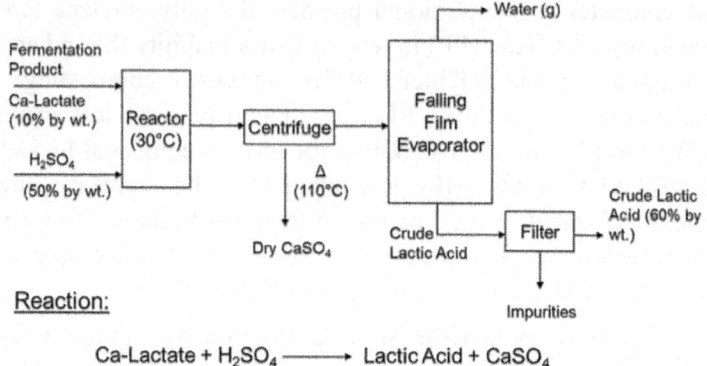

Reaction:

$$Ca\text{-}Lactate + H_2SO_4 \longrightarrow Lactic\,Acid + CaSO_4$$

Figure 1: Basic process flow diagram for the first stage of polymer grade lactic acid production.

In the second stage, the dilute lactic acid is sent to the counter-current reactive distillation column or the bubble column where it reacts with rising methanol vapors in the presence of concentrated H_2SO_4 to produce methyl lactate and water via the esterification reaction as depicted in Figure 2. Liquid methyl lactate flows to the bottom of the column while water and unreacted methanol vapors move up the column to the distillation section where they are separated. Liquid methyl lactate is then isolated by means of fractional distillation to separate out the methyl lactate by-products of fermentation impurity carboxylic acids to give highly pure methyl lactate (~98 wt.%).

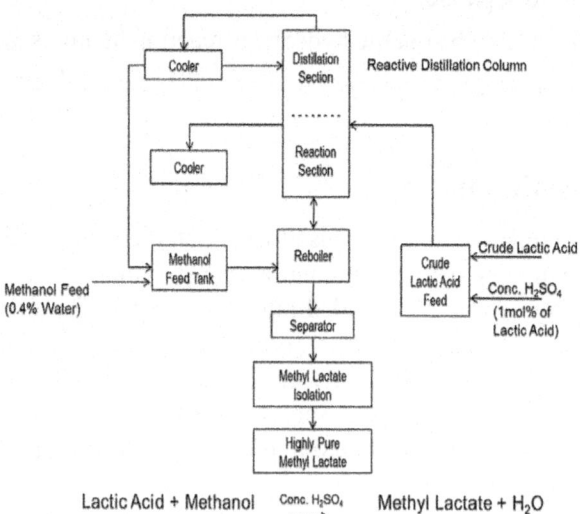

$$Lactic\,Acid + Methanol \xrightarrow{Conc.\,H_2SO_4} Methyl\,Lactate + H_2O$$

Figure 2: Basic process flow diagram for the second stage of polymer grade lactic acid production.

In this process, reactive distillation is the most crucial operation on which the entire process is dependent. A reactive distillation operation consists of simultaneous reaction and separation processes [14, 15]. In this case, methanol reacts with crude lactic acid to produce methyl lactate. As per the patented process, water and methanol are being simultaneously separated in the same column. However, while simulating this process it was observed that an additional separation column (Separat 4) was more effective for carrying out the separation of methanol and water than having a side-draw from the RadFrac column. It can therefore be said that the Bubble Column Reactor and Separat 4 together represent the Reactive Distillation Column. However, it should be noted that the more significant separation of methyl lactate and water + methanol is still occurring in the Bubble Column Reactor.

As the product (i.e. methyl lactate) is being continuously removed from the reactor the reaction equilibrium shifts towards the product side, thereby increasing maximum conversion according to Le-Chatelier's principle.

This is followed by the third stage which involves hydrolysis of pure methyl lactate to high purity lactic acid (Figure 3). The ingenuity of this stage lies in the fact that pure lactic acid is used as an auto-catalyst to avoid impurities. The use of an auto-catalyst facilitates achieving high lactic acid purity (99 wt. % on dry basis) and also increases the reaction rate. Also, methanol which is a by-product of the hydrolysis reaction is recycled back to the bubble column, thereby reducing the inventory cost and energy. The design objective for this simulation is to obtain a high yield of polymer grade (99 wt. % on dry basis) lactic acid as the final product. The process is optimized by varying the stream flow rates and design conditions to come up with the most energy efficient alternative for a given lactic acid output. The process model is created and simulated using Aspen Plus® version 8.2.

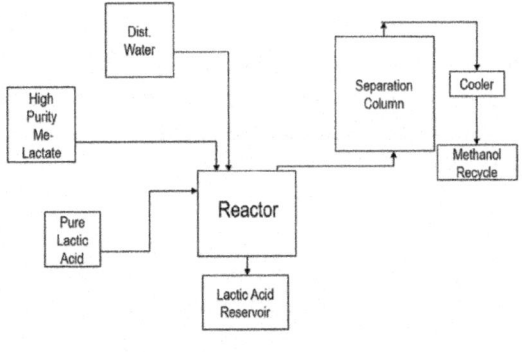

Methyl Lactate + H_2O $\xrightarrow{\Delta}$ Lactic Acid + Methanol

Figure 3: Basic process flow diagram for the third stage of polymer grade lactic acid production.

It should be noted that the process flow diagrams depicted in Figures 1, 2 and 3 have been retained as per the process described in the patent and should be used for reference purpose only. While performing the simulations certain changes have been made to the process for either simplicity or improved productivity. The final simulated process flow diagram is as per shown in Figure 4.

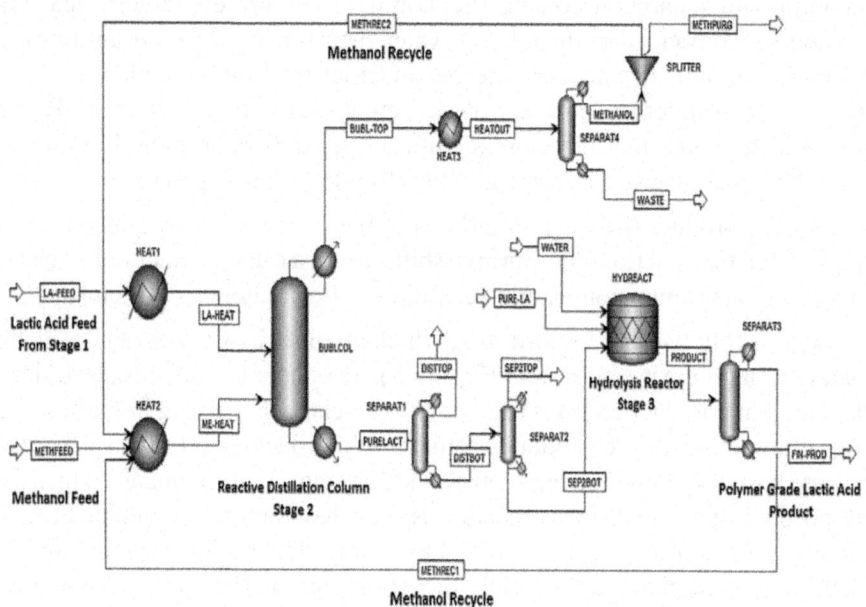

Figure 4: Process flow diagram in Aspen Plus for production of polymer grade lactic acid.

Design basis

The design basis was setup based on Example 6 (Control Example) and Example 10 (hydrolysis of methyl lactate to get pure lactic acid) from the patent [13]. The design specifications are listed in Table 2.

Table 2: Design basis obtained from patented process

Equipment	Parameter	Values
Bubble column reactor	Crude lactic acid + Impurities	1000 g/hr
	Lactic acid feed temperature	369 K
	Methanol + water	750 g/hr
	Methanol feed temperature	392 K
	Bottom section temp of bubble column	376 K
Separator	Highly pure methyl lactate	2500 g/hr
Hydrolysis reactor	Distilled water to reactor	2500 g/hr
	Pure lactic acid to reactor	500 g/hr
	Reactor temperature	373 K

Model Development

The following step-wise methodology was adopted while simulating the process.

Thermodynamic Model Selection

The majority of the process deals with highly polar components like lactic acid and methyl lactate, so the model must be based on the activity coefficient. In addition, lactic acid and methyl lactate both display non-ideal behavior which again indicates the requirement of an activity coefficient model. Therefore, initially the NRTL thermodynamic model was used as the thermodynamic property model for this process. But, the NRTL property model works well only for the vapor phase predictions of VLE. Hence, the Wilson-Hayden O'Connell property model was selected which provides accurate predictions for both phases in VLE.

Unit Operation Model selection

The first step in model development was to select the appropriate unit operation model based on the information available and the required output. For simulating the reactive distillation, RadFrac unit operation model is used. The RadFrac model can be used for rigorous fractionation, mainly for two or three phase vapor–liquid fractionation. Another advantage of RadFrac is the fact that it can handle chemical reactions, which is the case for reactive distillation column and also RadFrac can deal with strong liquid phase non-ideality. The hydrolysis reactor is modeled using the RStoich reactor. The main reason for selecting RStoich is that both conversion as well as stoichiometry was known for the reaction. Finally, Distl and Sep 2 model was used to simulate the two separators; the one used for isolating pure methyl lactate in stage 2 and the other used for separating pure lactic acid in stage 3.

PFD Generation

Based on the unit operation model selection, a basic process flow diagram (PFD) is generated in Aspen Plus® (Figure 4).

Sensitivity Analysis

Sensitivity Analysis is a technique available in the Aspen Plus software which can be used to ascertain the effects of certain parameters on significant process variables. It can therefore be used to obtain optimum process conditions. In this technique different process parameters can be varied independently or simultaneously to study their effect on process variables like mass/mole fraction of desired product (mass/mole purity). In this work, sensitivity analyses has been performed for several objective variables including bubble column reactor and separator parameters- number of stages, distillate to feed ratio, reflux ratio, feed stage; hydrolysis reactor parameters – temperature, pure lactic acid flow, distilled water flow; methanol recycle split ratio.

RESULTS AND DISCUSSION

Sensitivity analyses were carried out for several objective variables as shown in Table 3. As can be observed from the table, methyl lactate and lactic acid purity (mass fraction) are the main objective variables. The sensitivity analyses were performed in a systematic manner for each of the major equipment – Bubble Column Reactor, Hydrolysis Reactor, Lactic Acid Separator (Separator 3), Methyl Lactate Separator (Separator 1), Methanol Recycle Splitter, Heater 1, and Heater 2. The process variables were varied to study their effect on objective variables. A summary of the sensitivity analyses conducted and their results is illustrated in Table 3. The detailed tabulated results of the sensitivity analyses are provided in Additional file 1: Table S1-S6.

Table 3: Summary of sensitivity analyses performed

No.	Equipment	Process variables	Objective variables (Sensitivity)	Base case process variable	Process variable variation range	Objective variable variation range	Optimum process variable value
1	Lactic acid separator (Separator 3)	No. of stages	Lactic acid purity (Mass fraction)	9	9 - 11	0.99 – 0.99	9
2		Distillate to feed		*0.5*	*0.5 - 0.7*	*0.83 – 0.99*	*0.65*
3		Reflux ratio		1	1 - 1.3	0.99 – 0.99	1
4		Feed location		5	4 - 5	0.99 – 0.99	5
5	Hydrolysis reactor	Temperature (K)		373	313 - 413	0.62 – 0.62	373
6		Pure lactic acid flow (kg/hr)		0.015	0.01 – 0.05	0.62 – 0.62	0.01
7		Distilled water flow (kg/hr)		2	1.8 – 2.2	0.65 – 0.59	2
8	Bubble column reactor	No. of stages	Methyl lactate purity (Mass fraction)	17	5 - 17	0.63 – 0.64	17
9		Distillate to feed		*0.7*	*0.4 – 0.81*	*0.47 - 0.74*	*0.8*
10		Reflux ratio		2	1 - 4	0.64 – 0.64	1
11		Methanol feed stage		*15*	*13 - 17*	*0.59 – 0.64*	*15*
12	Heater 1	Temperature (K)		376	323 - 413	0.64 – 0.64	376
13	Heater 2	Temperature (K)		392	343 - 433	0.64 – 0.64	392
14	Splitter	Methanol recycle split fraction		1	0.15 - 1	0.74 – 0.64	1
15	Methyl lactate separator (Separator 1)	No. of stages		10	10 - 12	0.87 – 0.87	10
16		Distillate to feed		*0.6*	*0.1 – 0.6*	*0.67 – 0.87*	*0.6*
17		Reflux ratio		1.5	1 - 3	0.87 – 0.87	1
18		Feed location		5	3 - 8	0.86 – 0.87	6

Note: Italicized numbers depict variables with significant variations over the range.

Using the data from sensitivity studies, optimum process conditions were determined for each major unit. The optimum was based on achieving the maximum purity for methyl lactate and lactic acid with minimum chemical inventory. In case of temperatures, if no significant change in purity was observed, the base case values from the patent have been used. In all other cases where no significant change in purity was observed, process conditions understandably leading to least costs (capital and operating) and energy consumption, were used.

It can be noted from Table 3 that the distillate to feed ratio has a significant effect on purity in the bubble column reactor as well as both the separators (Separator 1 and Separator 3). On the other hand, the reflux ratio and number of stages do not have any significant change on purity in any of the three units. Purity is sensitive towards methanol feed location in the bubble column reactor but is not affected by the feed location in either of the two separators.

It was observed that methyl lactate purity increases with an increase in methanol feed stage till stage no. 15 after which it marginally decreases (Figure 5). This observation was expected because as the methanol feed stage increases (column stage number increases from top to bottom), the contact time for methanol and lactic acid increases, thereby increasing mass transfer, subsequently resulting in higher purity of methyl lactate. Along with the sensitivity analysis, the bubble column temperature profile was obtained. The study of profile variations helped in determining the optimum feed stage for the methanol feed stream (Figure 6).

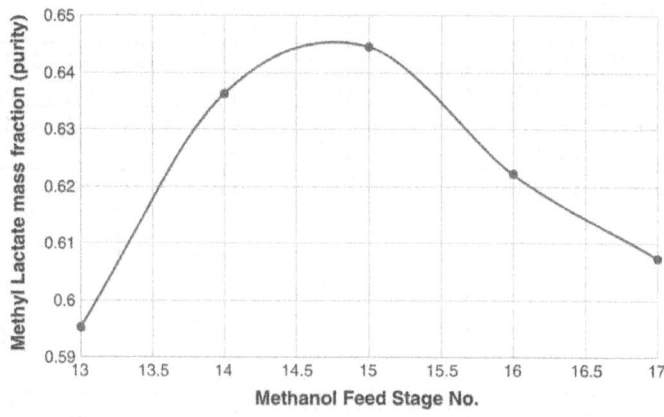

Figure 5: Sensitivity analysis results illustrating the effect of methanol feed stage variance in the bubble column reactor.

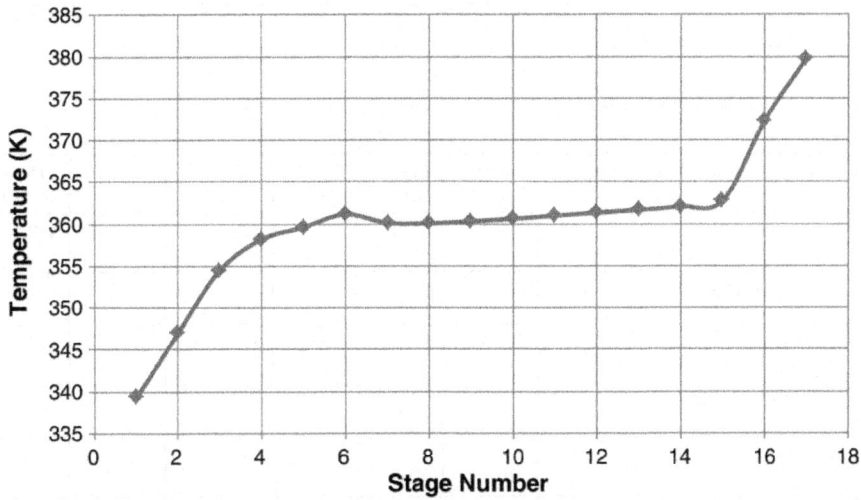

Figure 6: Temperature profile for the bubble column reactor (base case).

Stream results tables listing important streams with their process parameters for both the base case and optimized case have been illustrated in Tables 4 and 5. Comparing the two tables the improvements in product purity in the optimum case becomes evident. Note that the stream names used are as specified in the process flow diagram.

Table 4: Stream results table (Base case)

Parameter/ stream →↓	LA-FEED	METH-FEED	PURE-LACT	DIST-BOT	SEP-2BOT	ME-THREC1	PROD-UCT	FIN-PROD
Temperature (K)	298.1	298.1	379	393.4	393.4	343.2	373.1	413.7
Pressure (atm.)	1	1	1	1	1	1	1	1
Mass Fflow (kg/hr)	10.5	7.5	11.29	8.171	6.921	2.445	8.936	6.491
Mass fraction (Mass purity)								
Lactic acid	0.571		0.001	0.001		trace	0.603	*0.83*
Methyl lactate			0.613	*0.845*	*0.997*	147 PPM	0.077	0.106
Methanol		0.996	0.003	62 PPB		0.782	0.214	165 PPM
Water	0.381	0.004	0.339	0.093	0.003	0.218	0.106	0.063
Others	0.05		0.045	0.059				

Note: 1) Italicized numbers depict variables with significant variations over the range.

2) Listed row headers depict streams and column headers depict parameters (As shown by arrows).

Table 5: Stream results table (Optimized case)

Parameter/ stream →↓	LA-FEED	METH-FEED	PURE-LACT	DIST-BOT	SEP-2BOT	ME-THREC1	PROD-UCT	FIN-PROD
Temperature (K)	298.1	298.1	382.4	414.6	414.6	346.6	373.1	489.8
Pressure (atm)	1	1	1	1	1	1	1	1
Mass flow (kg/hr)	10.5	7.5	11.056	8.625	8.072	4.129	10.122	5.993
Mass fraction (Mass purity)								
Lactic acid	0.571		730 PPM	935 PPM		0.079	0.624	*1*
Methyl lactate			0.733	*0.933*	*0.997*	0.195	0.08	41 PPM
Methanol		0.996	0.005	302 PPB		0.54	0.22	4 PPB
Water		0.381	0.004	0.216	0.009	0.003	0.186	0.076	57 PPB
Others	0.05		0.045	0.059				

Note: 1) Italicized numbers depict variables with significant variations over the range.

2) Listed row headers depict streams and column headers depict parameters (As shown by arrows).

CONCLUSION

A process flow diagram for the production of polymer grade lactic acid was developed using Aspen Plus®. The process was successfully simulated and converged results were obtained. An extensive sensitivity analysis was conducted on each of the key pieces of equipment to obtain high purity methyl lactate and subsequently pure lactic acid. The process conditions were subsequently optimized based on these results. The desired purity of polymer grade lactic acid (99 wt. %, dry basis) was obtained. Also, the purity of methyl lactate was above the desired percentage of 98.5 wt. %.

ACKNOWLEDGEMENT

The authors would like to express their sincere acknowledgments to Dr. Sayeed Mohammad, School of Chemical Engineering, Oklahoma State University for providing timely assistance during this work.

ELECTRONIC SUPPLEMENTARY MATERIAL

Table S1: Sensitivity analyses results over **Separator 3** for **Lactic acid (LA) mass fraction** by varying Distillate to feed ratio, Reflux Ratio, Feed Location and No. of Stages

Sensitivity Analyses - Separator 3 (Aspen Model: Distl)							
Distillate to Feed(D:F)		**Reflux Ratio (RR)**		**Feed Location (F-LOC)**		**No. of Stages (NSTAGE)**	
D:F	LA Mass Frac	RR	LA Mass Frac	F-LOC	LA Mass Frac	N-STAGE	LA Mass Frac
0.5	0.83042	1	0.99996	4	0.99998	9	0.99996
0.55	0.85155	1.1	0.99997	5	0.99996	10	0.99999
0.6	0.87376	1.2	0.99997			11	1.00000
0.65	0.99996	1.3	0.99997				
0.7	0.99997						

Table S2: Sensitivity analyses results over **Hydrolysis Reactor** for **Lactic acid (LA) mass fraction** by varying Reactor Temperature, Pure Lactic Acid Inlet Flow Rate and Distilled Water Flow rate

Sensitivity Analyses - Hydrolysis Reactor (Aspen Model: RStoich)					
Reactor Temperature		**Pure Lactic Acid Inlet Flow Rate**		**Water Flow Rate**	
Temperature (K)	LA Mass Frac	Pure Lactic Acid Mass Flow Rate (KG/HR)	LA Mass Frac	Water Mass Flow Rate (KG/HR)	LA Mass Frac
313	0.623120392	0	0.62208008	1.8	0.650124349
323	0.623121749	0.01	0.622775545	1.85	0.643813146
333	0.62312211	0.015	0.623122035	1.9	0.637221173
343	0.623121704	0.02	0.623468003	1.95	0.630328981
353	0.623121176	0.03	0.624155377	2	0.623121728
363	0.623121734	0.04	0.624838559	2.05	0.615573215
373	0.623122075	0.05	0.625517586	2.1	0.607667016

373.15	0.623121998			2.15	0.599386385
383	0.623122425			2.2	0.590881851
393	0.623122202				
403	0.623122008				
413	0.623122056				

Table S3: Sensitivity analyses results over **Bubble Column Reactor** for **Methyl Lactate (Me-La) mass fraction** by varying No. of Stages, Distillate to feed ratio, Reflux Ratio and Methanol Feed Location

Sensitivity Analyses - Bubble Column Reactor (Aspen Model: RadFrac)							
No. of Stages (NSTAGE)		**Distillate to Feed(D:F)**		**Reflux Ratio (RR)**		**Methanol Feed Location (F-LOC)**	
NSTAGE	Me-La Mass Fraction	D:F	Me-La Mass Fraction	RR	Me-La Mass Fraction	F-LOC	Me-La Mass Fraction
5	0.6302035	0.4	0.4728521	1	0.6412163	13	0.5951648
6	0.6305344	0.45	0.5006606	1.2	0.6438805	14	0.6362666
7	0.6282199	0.5	0.5310520	1.4	0.6445551	15	0.6445160
8	0.6275278	0.55	0.5594982	1.6	0.6447084	16	0.6221627
9	0.6273122	0.6	0.5841040	1.8	0.6446682	17	0.6074414
10	0.6272492	0.65	0.6118929	2	0.6445105		
11	0.6272341	0.7	0.6445148	2.2	0.6443002		
12	0.6272271	0.75	0.6837251	2.4	0.6440343		
13	0.6272245	0.8	0.7310741	2.6	0.6437424		
14	0.6272229	0.81	0.7403386	2.8	0.6434277		
15	0.6272224			3	0.6430896		
16	0.6400223			3.2	0.6427302		
17	0.6445154			3.4	0.6423586		
				3.6	0.6419719		
				3.8	0.6415639		

				4	0.6411479		

Table S4: Sensitivity analyses results over **Bubble Column Reactor** for **Methyl Lactate (Me-La) mass fraction** by varying Heater 1 (Lactic Acid Feed) Temperature, Heater 2 (Methanol Feed) Temperature

Sensitivity Analyses - Bubble Column Reactor (Aspen Model: RadFrac)			
Heater 1 (Lactic Acid Feed) Temperature		Heater 2 (Methanol Feed) Temperature	
Heat 1 Temp (K)	Me-La Mass Fraction	Heat 2 Temp (K)	Me-La Mass Fraction
323	0.6444547	343	0.6444983
333	0.6444665	353	0.6445105
343	0.6444773	363	0.6445093
353	0.6444890	373	0.6445103
363	0.6444987	383	0.6445117
373	0.6445102	392.15	0.6445146
376.15	0.6445141	393	0.6445130
383	0.6445829	403	0.6445142
393	0.6446868	413	0.6445152
403	0.6447078	423	0.6445162
413	0.6447132	433	0.6445171

Table S5: Sensitivity analyses results over **Splitter** for **Methyl Lactate (Me-La) mass fraction** by varying methanol split fraction

Sensitivity Analyses - Splitter (Aspen Model: FSplit)	
Methanol Split fraction	Me-La Mass Fraction
0.15	0.7414326
0.2	0.7380813
0.25	0.7340047
0.3	0.7294677
0.35	0.7246340
0.4	0.7195749
0.45	0.7143304
0.5	0.7089075
0.55	0.7033045

0.6	0.6975320
0.65	0.6915837
0.7	0.6854536
0.75	0.6791417
0.8	0.6726302
0.85	0.6659218
0.9	0.6590059
0.95	0.6518723
1	0.6445289

Table S6: Sensitivity analyses results over **Separator 1** for **Methyl Lactate (Me-La) mass fraction** by varying Distillate to feed ratio, Reflux Ratio, Feed Location and No. of Stages

Sensitivity Analyses - Separator 1 (Aspen Model: Distl)							
Distillate to Feed(D:F)		Reflux Ratio (RR)		Feed Location (F-LOC)		No. of Stages (NSTAGE)	
D:F	Me-La Mass Frac	RR	Me-La Mass Frac	F-LOC	Me-La Mass Frac	N-STAGE	Me-La Mass Frac
0.1	0.6749167	1	0.8703432	3	0.8608662	10	0.8718628
0.15	0.6902378	1.1	0.8708207	4	0.8691948	11	0.8719710
0.2	0.7063280	1.2	0.8711905	5	0.8718601	12	0.8720475
0.25	0.7232820	1.3	0.8714737	6	0.8727564		
0.3	0.7411614	1.4	0.8716885	7	0.8730763		
0.35	0.7600293	1.5	0.8718580	8	0.8731814		
0.4	0.7799579	1.6	0.8720016				
0.45	0.8010205	1.7	0.8721181				
0.5	0.8232983	1.8	0.8722216				
0.55	0.8468806	1.9	0.8723029				
0.6	0.8718575	2	0.8723737				
		2.1	0.8724322				
		2.2	0.8724850				
		2.3	0.8725238				

		2.4	0.8725688				
		2.5	0.8726009				
		2.6	0.8726375				
		2.7	0.8726633				
		2.8	0.8726778				
		2.9	0.8727005				
		3	0.8727211				

AUTHORS' CONTRIBUTIONS

SSB performed process simulations and optimization studies for the process presented in the paper. SSB drafted the manuscript in collaboration with CPA. CPA provided consultation regarding the process simulations and optimization. CPA drafted the manuscript in collaboration with SSB. KAH provided consultation regarding the process simulations and optimization. KAH reviewed the manuscript. All authors read and approved the final manuscript.

REFERENCES

1. Datta R, Henry M: Lactic acid: recent advances in products, processes and technologies — a review. *J Chem Technol Biotechnol*2006,81(7):1119–1129. doi:10.1002/jctb.1486

2. Guilherme A, Silveira M, Fontes C, Rodrigues S, Fernandes F: Modeling and optimization of lactic acid production using cashew apple juice as substrate. *Food Bioprocess Technol* 2012,5(8):3151–3158. doi:10.1007/s11947-011-0670-z

3. Södergård A, Stolt M: Industrial Production of High Molecular Weight Poly (Lactic Acid). In *Poly (Lactic Acid): Synthesis, Structures, Properties, Processing, and Applications*. Edited by: Auras R, Lim L-T, Selke SEM, Tsuji H. Hoboken, NJ, USA: John Wiley & Sons, Inc; 2010. doi:10.1002/9780470649848.ch3

4. Garlotta D: A literature review of poly(lactic acid). *J Polym Environ* 2001,9(2):63–84. doi:10.1023/A:1020200822435.

5. Qin J, Zhao B, Wang X, Wang L, Yu B, Ma Y, Xu P: Non-sterilized fermentative production of polymer-grade L-lactic acid by a newly isolated thermophilic strain Bacillussp. *PLoS ONE* 2009,4(2):e4359. doi:10.1371/journal.pone.0004359

6. John R, Nampoothiri K, Pandey A: Fermentative production of lactic acid from biomass: an overview on process developments and

future perspectives. *Appl Microbiol Biotechnol* 2007,74(3):524–534. doi:10.1007/s00253–006–0779–6

7.   Mehta R, Kumar V, Bhunia H, Upadhyay S: Synthesis of poly(lactic acid): a review. *J Macromol Sci Part C* 2005,45(4):325–349. doi:10.1080/15321790500304148.

8.   Tolinski M: *Plastics and Sustainability - Towards a Peaceful Coexistence between Bio-Based and Fossil Fuel-Based Plastics*. Salem, Massachussetts: Wiley-Scrivener; 2012.

9.   Tabone M, Cregg J, Beckman E, Landis A: Sustainability metrics: life cycle assessment and green design in polymers. *Environ Sci Technol* 2010,44(21):8264–8269.

10.   Henry W: *Purifying hydroxyl-aliphatic acids*. New York; 1943. [*United States Patent 2334524*]

11.   Schopmeyer H, Arnold C: *Lactic Acid Purification*. Maine: United States Patent 2350370; 1944.

12.   Weisberg S, Stimpson E: *Preparation of Lactic Acid*. Baltimore, Maryland; 1942. [*United States Patent 2290926*]

13.   Barve P, Kulkarni B, Nene S, Shinde R, Gupte M, Joshi C, Thite G, Chavan V, Deshpande T: *Process for Preparing L(+)-Lactic Acid*. Maharashtra, India: United States Patent 7820859 B2; 2010.

14.   Stichlmair J, Frey T: Reactive distillation processes. *Chem Eng Technol* 1999,22(2):95–103. doi:10.1002/(SICI)1521–4125(199902)22:2<95::AID-CEAT95>3.0.CO;2-#

15.   Hauan S, Lien K: A phenomena based design approach to reactive distillation. *Chem Eng Res Des* 1998,76(3):396–407.

Chapter 9

SCENARIO ANALYSIS OF A BIOETHANOL FUELED HYBRID POWER GENERATION SYSTEM

Wei Wu[1], Yuan-Tai Hsu[2] and Po Chih Kuo[1]

[1]Department of Chemical Engineering, National Cheng Kung University, Tainan 70101, Taiwan

[2]National Yunlin University of Science and Technology, Douliou, Yunlin 64002, Taiwan.

ABSTRACT

Background

Regarding most of FC/PV/Battery based hybrid power generation systems, the photovoltaic (PV) power usually dominates the main power supply and the water electrolyzer is used to produce hydrogen. The energy efficiencies of these hybrid power generation (HPG) systems are usually low due to the low conversion efficiency of PV cell and the extra power consumption to hydrogen production. To reduce the electricity demand by the PV system and improve the energy efficiency of hydrogen production unit, a scenario-based design of the HPG system is necessary.

Result

This paper proposes an EFC/PV/Battery based hybrid power generation system to meet 24-hour power demand. An ethanol-fueled fuel cell (EFC) power generator not only dominates the main power supply, but also a combination of the stand-alone EtOH-to-H2 processor and PEMFC can ensure higher energy efficiency. A PV system is treated as an auxiliary power generator which can reduce the (bio)ethanol consumption. A backup battery not only stores excess power from PV or EFC, but also it can precisely satisfy the power demand gap. Finally, scenario analysis of the hybrid power generation (HPG) system in regard to the hybrid power dispatch and energy efficiency is addressed.

Conclusion

An optimized fuel processing unit using ethanol fuel can produce high-purity hydrogen. The simulation shows that the stand-alone EtOH-to-H2 processor not only guarantee the high energy efficiency, but also it can continuously produce hydrogen if the fuel is enough. According to scenarios for the daily operation of the HPG system, the EFC power dominates the power supply during the night, the PV system dominates the power supply during the day and the backup battery aims to instantly compensate the power gap and store the excess power from PV or EFC. According to the hybrid power dispatch, the distribution of the HPG system efficiency is specified.

BACKGROUND

Recently the interest in renewable and sustainable power generation techniques such as photovoltaic, wind, and fuel cell powered systems has grown. The off-grid hybrid power system is increasingly popular for remote area power generations, even though the photovoltaic (PV) or wind system is highly dependent on weather conditions and locations. Muselli et al. [1] developed a stand-alone hybrid power system which was composed of the PV panels, an electrical generator using fossil fuels, and battery storage, and Elhadidy [2] studied the feasibility of using wind/solar/diesel energy conversion systems to meet the power demand of the specific community in Saudi Arabia. Notably, a gasoline or diesel engine is usually treated as the backup power generator.

The PV system generates electricity during daylight hours, but the fuel cell system produces electricity as long as hydrogen is supplied. To address the clean power generation system, Hwang et al. [3] proposed a solar/fuel cell hybrid power system to meet the daily load demand of a typical family in Taiwan, Uzunoglu et al. [4] investigated dynamic models and simulation of this kind of hybrid power generation systems with appropriate control strategies, and Zervas et al. [5] provided an optimal decision framework for managing the hybrid energy systems. Similarly, these hybrid systems all involve the water electrolyzer for producing hydrogen. The electrolyzer is simpler and does not require complex units, but there is no solution to reduce its large electricity consumption. For most of hybrid systems, the PV panel is larger so as to carry out water electrolysis by using the excess power. Therefore, the overall energy efficiency is quite low [6].

The hydrogen fuel cells have been considered the most promising for automotive application, but the guidelines for the hydrogen storage, safer operation and transportation are quite rigorous [7]. The fossil fuel processing unit could produce the high purity of hydrogen [8, 9], but the external energy supply and large carbon emissions are not avoided. If biomass can replace fossil fuels as the feedstock, the net-zero greenhouse gas (GHG) emissions of hydrogen production systems can be achieved. Liquefied ethanol is a popular feedstock because it is easily made by fermentation, no compressed storage is needed, and lower operating temperatures are required for reforming processes [10,11].

In this paper, an ethanol fueled HPG system is proposed. The modeling, optimization and design of the EtOH-to-H_2 processor is achieved by using Aspen Plus. The scenario analysis of the HPG system in regard to the daily load demand of a typical family in Taiwan [3] is investigated in Matlab. Since the EFC power unit cannot rapidly meet the load demand due to complicated chemical reactions in the ethanol reforming process, the lithium-ion battery is considered to meet the power gap when the excess power from PEMFC and PV units have been completely stored in the battery in advance. According to prescribed hybrid power dispatch, it is verified that the energy efficiency of the proposed HPG system is superior to the other renewable hybrid energy system.

Hybrid Power Generation System

The ethanol-fueled HPG system is shown in Figure 1. Both PV generator and PEMFC stack are treated as power sources. The lithium-ion battery is denoted as an electricity storage/instant supply device. The stand-alone EtOH-to-H_2 processor in place of the devices of the hydrogen storage tank and water electrolyzer aims to produce high purity hydrogen. Other devices, including converters/inverters and power management (DC bus), are used to meet the voltage specification.

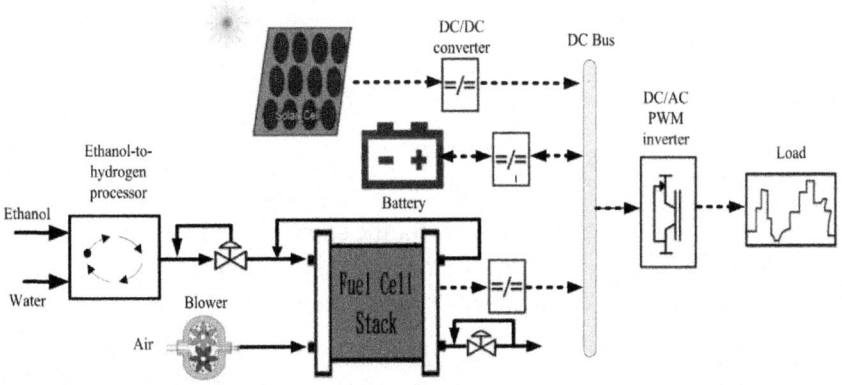

Figure 1: Hybrid power generation system.

EtOH-to-H$_2$ Processor

The EtOH-to-H$_2$ processor shown in Figure 2 is considered as a stand-alone hydrogen production system, which is composed of an ethanol steam reforming (ESR) reactor, a low-temperature water-gas-shift (LTWGS) reactor, the pressure swing adsorption (PSA) unit, and a burner. The simple heat integration is adopted to improve the water heat recovery; (i) the waste heat generated from the burner is applied to preheat the mixing feed flow using a heat exchanger (HEX1) and then meets the heat demand of the reforming reactions in the ESR reactor; (ii) the inlet water flow is preheated by a heat exchanger (HEX2) to increase the feed temperature; (iii) the temperature of flue gas discharged from the heating jacket of the ESR reactor is expected to be low.

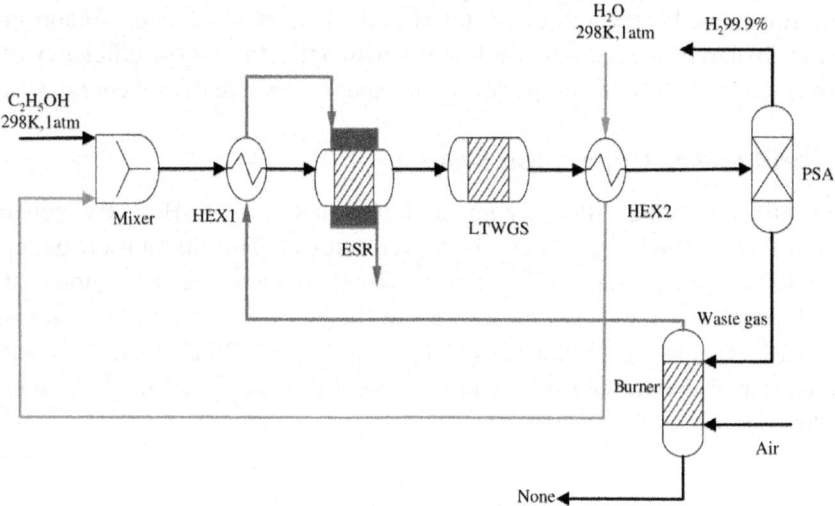

Figure 2: Stand-alone EtOH-to-H$_2$ processor.

ESR Reactor

According to the kinetic experiments for the ethanol steam reforming reactions over Co$_3$O$_4$-ZnO catalyst at atmospheric pressure and temperature in the 600-800 K range [12], first the reactions of ESR are shown as:

$$C_2H_5OH \rightarrow CH_3CHO + H_2 \ (r_1), \Delta H_1^0 = 71 \text{ kJ mol}^{-1} \tag{1}$$

$$C_2H_5OH \rightarrow CO + CH_4 + H_2 \ (r_2), \Delta H_2^0 = 52.9 \text{ kJ mol}^{-1} \tag{2}$$

$$CO + H_2O \leftrightarrow CO_2 + H_2 \ (r_3), \Delta H_3^0 = -34.5 \text{ kJ mol}^{-1} \tag{3}$$

$$CH_3CHO + 3H_2O \rightarrow 2CO_2 + 5H_2 \ (r_4), \Delta H_4^0 = 127.4 \text{ kJ mol}^{-1} \tag{4}$$

Second, the kinetic models of four reactions (r_1, ..., r_4) are expressed by power-law rate expressions [12]:

$$r_1 = 2.1 \times 10^4 \exp\left(-\frac{70}{R}\left(\frac{1}{T_{ESR}}-\frac{1}{773}\right)\right)P_{C_2H_5OH} \tag{5}$$

$$r_2 = 2.0 \times 10^3 \exp\left(-\frac{130}{R}\left(\frac{1}{T_{ESR}}-\frac{1}{773}\right)\right)P_{C_2H_5OH} \tag{6}$$

$$r_3 = 1.9 \times 10^4 \exp\left(-\frac{70}{R}\left(\frac{1}{T_{ESR}}-\frac{1}{773}\right)\right)$$
$$\times \left(P_{CO}P_{H_2O}-\frac{P_{C_2O}P_{H_2}}{\exp\left(\frac{4577.8}{T_{ESR}}-4.33\right)}\right) \tag{7}$$

$$r_4 = 2.0 \times 10^5 \exp\left(-\frac{98}{R}\left(\frac{1}{T_{ESR}}-\frac{1}{773}\right)\right)\left(P_{CH_3CHO}P_{H_2O}^3\right) \tag{8}$$

where T_{ESR} is the temperature of the ESR reactor, P_i (i=C$_2$H$_5$OH, CO, H$_2$O, ...) is corresponding partial pressure, and R is the universal gas constant. Moreover, the one-dimensional, pseudo-homogeneous model is built by the Aspen Plus. The assumptions for the steady-state simulation of the ESR reactor include gas phase reactions, plug-flow reactor, and no pressure drop. The thermodynamic properties of some species are evaluated by using the Peng-Robinson equation of state. Other parameters include the fluid density=0.02 kmol m^{-3}, catalyst weight=5 kg, and sizes of the reactor with the diameter and length of the reactor D=0.5 m and L=1.5 m.

LTWGS Reactor

To reduce CO as well as enhance hydrogen yield, the LTWGS reactor could be directly connected to the ESR reactor, where the reaction scheme is:

$$CO + H_2O \xrightleftharpoons{Cat} CO_2 + 2H_2, \quad \Delta\hat{H}_R^o = -41.15 \text{ kJ mol}^{-1} \tag{9}$$

For a moderately exothermic reaction with Sud-Chemie commercial available catalysts, we assume that the reactor is adiabatic and the operating temperature is between 400 K and 600 K. Moreover, the kinetic model is described by [13],

$$r_W = 82.2 \exp\left(-\frac{47400}{RT_{WGS}}\right)\left(P_{CO}P_{H_2O}-\frac{P_{CO_2}P_{H_2}}{K_{WGS}}\right) \tag{10}$$

where T_{WGS} is the temperature of the WGS reactor and K_{WGS} is the equilibrium constant of the WGS reaction shown by

$$\ln(K_{WGS}) = \frac{5693.5}{T_{WGS}} + 1.077 \ln(T_{WGS}) + 5.44 \times 10^{-4} T_{WGS}$$
$$-1.125 \times 10^{-7} T^2_{WGS} - \frac{49170}{T^2_{WGS}} - 13.148$$

(11)

Burner

Regarding the design of burner, we assume that (i) H_2 is completely separated by PSA, (ii) all components of the burner are completely burned, and (iii) its Aspen module is considered as an adiabatic and stoichiometric reactor. The combustion reaction in the burner is shown as follows:

$$C_2H_5OH + 3O_2 \rightarrow 2CO_2 + 3H_2O, \Delta \hat{H}^o_{b,EtOH} = -1279 \text{ kJ mol}^{-1}$$

(12)

$$CH_4 + 2O_2 \rightarrow CO_2 + 2H_2O, \Delta \hat{H}^o_{b,CH_4} = -803.1 \text{ kJ mol}^{-1}$$

(13)

$$H_2 + \frac{1}{2}O_2 \rightarrow H_2O, \Delta \hat{H}^o_{b,H_2} = -242 \text{ kJ mol}^{-1}$$

(14)

$$CO + \frac{1}{2}O_2 \rightarrow CO_2, \Delta \hat{H}^o_{b,CO} = -283 \text{ kJ mol}^{-1}$$

(15)

Sensitivity anAlysis

Regarding the effect of fuel utilization, the ratio of hydrogen to ethanol (H_2/C_2H_5OH) is specified by adjusting ethanol flow or water flow in the feed.

1.   The profiles of H_2/C_2H_5OH vs. $(S/C)|_{H_2O}$ at different $T_{ESR,in}$ is depicted in Figure 3(a). It shows that the maximum hydrogen yield can be achieved by adjusting the water flow at constant ethanol flow. The profiles of H_2/C_2H_5OH vs. $(S/C)|_{EtOH}$ at different $T_{ESR,in}$ is depicted in Figure 3(b). Similarly, it shows that the maximum hydrogen yield is obtained by adjusting the ethanol flow at constant water flow. The corresponding operating ranges are bounded by $2 \leq (S/C)|_{EtOH,H_2O} \leq 2.5$ and $600K \leq T_{ESR,in} \leq 800K$. Notably, the manipulation of ethanol flow can ensure the larger hydrogen yield than the water flow at the same $T_{ESR,in}$, and the increase of $T_{ESR,in}$ cannot induce the high hydrogen yield. According the heat recovery design, the waste gas temperature at the outlet of heating jacket decreases when the preheated temperature of ESR increases. The low reactor temperature would reduce the conversion of ethanol reforming reactions.

2.   According to the sensitivity analysis by adjusting one variable $(S/C)|_{EtOH}$, the optimization tool of Aspen Plus is employed to determine the optimal operating conditions, $(S/C)|_{EtOH} = \frac{27.2 \text{ kgmol/hr}}{12.7 \text{ kgmol/hr}} = 2.14$ at $T_{ESR,in} = 600$ K. The steady-state simulation of the system at optimal conditions is shown

in Figure   4, where the flowrate of hydrogen product can achieve 55.62 kgmol/hr, i.e. H_2/C_2H_5OH =4.38 and the ethanol conversion can achieve 95%.

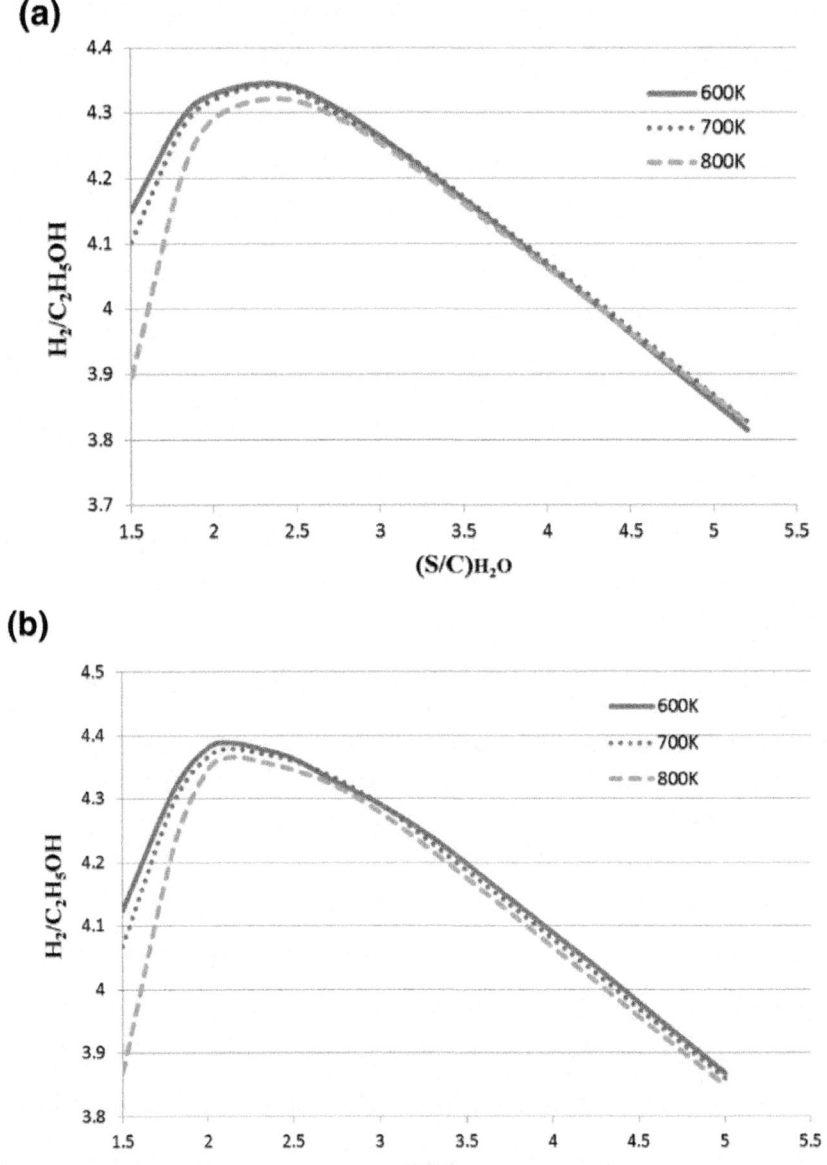

Figure 3: Fuel utilization ratio of ethanol-to hydrogen processor at different $T_{ESR,in}$ by adjusting (a) $2 \leq (S/C)|_{EtOH,H_2O} \leq 2.5$ $(S/C)|_{H2O}$ (b)$(S/C)|_{EtOH}$.

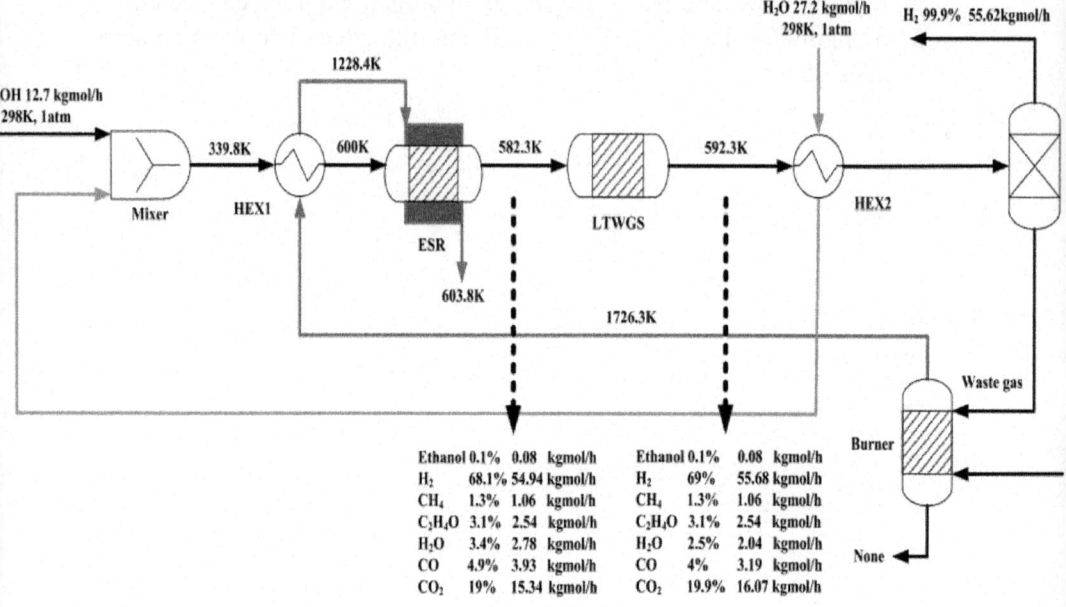

Figure 4: Optimization and simulation of EtOH-to-hydrogen processor.

PEM Fuel Cell System

A Ballard 5 kW PEMFC stack system consists of 30 cells connected in series to provide output current up to 100 A, which is directly connected to an ethanol-to-hydrogen processor. The high purity of hydrogen produced from the processor is fed to the anode and the excess hydrogen needs to be recirculated. Referring to an empirical fuel cell stack model by [14, 15], the output voltage of a single fuel cell (V_{fc}) is expressed by

$$V_{fc} = E - V_{act} - V_{ohm} \tag{16}$$

The open circuit cell potential E via the Nernst equation is described by

$$E = 1.229 - 8.5 \times 10^{-4}(T_{fc} - 298.15) + \frac{RT_{fc}}{2F} \ln\left[P_{H_2}(P_{O_2})^{0.5}\right] \tag{17}$$

The activation overvoltage V_{act} is described by a first-order dynamic with the effects of double layer capacitance charging at the electrode-electrolyte interfaces

$$\frac{dV_{act}}{dt} = \frac{i_{fc}}{C_{dl}} + \frac{i_{fc}E_{act}}{V_{act}C_{dl}} \tag{18}$$

where the activation drop E_{act} is defined by

$$E_{act} = \beta_1 + \beta_2 T_{fc} + \beta_3 T_{fc} \ln(C_{O_2}) + \beta_4 T_{fc} \ln(i_c) \tag{19}$$

and the parametric coefficients $\beta_1, ..., \beta_4$ are expressed by

$\beta_1 = -0.948$

$\beta_2 = 0.00286 + 0.0002 \ln(A_{fc}) + 4.3 \times 10^{-5} \ln(C_{H_2})$

$\beta_3 = 7.6 \times 10^{-5}$

$\beta_4 = -1.93 \times 10^{-4} \tag{20}$

and

$$C_{O_2} = 1.97 \times 10^{-7} P_{O_2} \exp\left(\frac{498}{T_{fc}}\right)$$

$$C_{H_2} = 9.174 \times 10^{-7} P_{H_2} \exp\left(\frac{-77}{T_{fc}}\right) \tag{21}$$

The ohmic overvoltage is given by:

$$V_{ohm} = \frac{l_m}{A_{fc}} r_M i_{fc} \tag{22}$$

where the membrane resistivity r_M is described by

$$r_M = \frac{181.6\left[1 + 0.03\left(\frac{i_{fc}}{A_{fc}}\right) + 0.062\left(\frac{T_{fc}}{303}\right)^2 \left(\frac{i_{fc}}{A_{fc}}\right)^{2.5}\right]}{\left[11.866 - 3\left(\frac{i_{fc}}{A_{fc}}\right)\right] \exp\left[4.18\left(\frac{T_{fc}-303}{T_{fc}}\right)\right]} \tag{23}$$

Above of all parameters for modeling the PEMFC system have been defined in the nomenclature section. Based on the ideal gas law and the principle of mole conservation, the partial pressures of hydrogen (P_{H_2}) and oxygen (P_{O_2}) associated with reactant flow rates at the anode and cathode along with the cell current are modeled as.

$$\frac{dp_{H_2}}{dt} = \frac{RT_{fc}}{V_{an}}\left[\dot{m}_{H_2,in} - k_{an}\left(p_{H_2} - p_{H_2,in}\right) - \frac{15}{F} i_{fc}\right] \tag{24}$$

$$\frac{dp_{O_2}}{dt} = \frac{RT_{fc}}{V_{ca}}\left[\dot{m}_{O_2,in} - k_{ca}\left(p_{O_2} - p_{BPR}\right) - \frac{7.5}{F} i_{fc}\right] \tag{25}$$

Notably, two pressure regulators with fixed parameters (k_{an}, k_{ca}) are added to manipulate the outlet hydrogen flow at the anode and the outlet oxygen flow at the cathode. The above model equations show that the fuel cell performance is affected by current density (i_{fc}), fuel cell temperature (T_{fc}), hydrogen and oxygen partial pressure. Referring parameter values of the fuel cell model in Table 1, Figure 5 shows that the maximum cell power output appears when the current is operated around 250A and the cell temperature is fixed at $T_{fc} = 75°C$. Moreover, the voltage and power of the stack is given by $V_{stack} = 30V_{fc}$; $P_{fc} = V_{stack} i_{fc} A_{fc}$.

Table 1: PEMFC Parameters

Parameter	Value
V_{an}	0.005 m³
k_{an}	0.065 mol s⁻¹ atm⁻¹
P_{BPR}	1 atm
V_{ca}	0.01 m³
k_{ca}	0.065 mol s⁻¹ atm⁻¹
F	96485 Cmol⁻¹
A_{fc}	232 cm²
l_m	178×10^{-4} cm
R	8.314 Jmol⁻¹K⁻¹
C_{dl}	0.035×232 F
λ_{O_2}	2

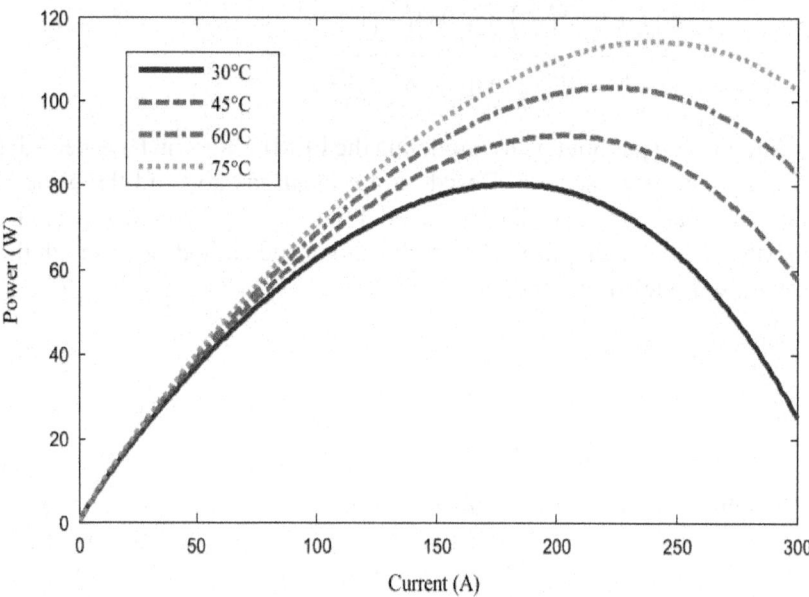

Figure 5: PEMFC power versus current at different T_{fc}.

Remark 1: To address the feasible manipulation of the fuel cell stack, two pressure regulators for manipulating the outlet flow rates of hydrogen and oxygen at the anode and cathode, respectively, are required. In addition, the oxygen starvation phenomenon should be avoided, i.e., the following inequality for oxygen excess ratio (λ_{O_2}),

$$\frac{\dot{m}_{O_2,in}}{8.75 i_{fc}/F} \geq \lambda_{O_2}$$

(26)

should be satisfied by increasing the inlet flow rate of oxygen ($\dot{m}_{O_2,in}$) or air flow [15].

Remark 2: The stack requires a circulating water system to keep the relative humidity of membrane as well as regulate the stack temperature. It is assumed that the stack system is isothermal, the relative humidity is kept at a desired level, and the hydrogen flow in the inlet of fuel cell ($\dot{m}_{H_2,in}$). In fact, the fuel processing delays, including reaction time and transportation delay exist such that the power gap between supply and demand is inevitable.

Photovoltaic Generator

A photovoltaic (PV) generator is composed of numbers of solar cells connected in series and parallel to provide the desired output terminal voltage and current. Using the Kirchoff's current law the terminal current I_s through the solar cell is expressed by [3].

$$I_s = I_{ph} - I_d - I_{sh}$$

(27)

The photocurrent I_{ph} is described as

$$I_{ph} = p_1 E_s \left[1 + p_2 (E_s - E_0) + p_3 (T_j - T_0) \right]$$

(28)

The diode loss current I_d is described as

$$I_d = p_4 T_j^3 \exp\left(-\frac{E_g}{\kappa T_j}\right) \left[\exp\left(\frac{e_0}{a_f N_s \kappa} \frac{V_s + R_s I_s}{T_j}\right) - 1 \right]$$

(29)

The shunt current I_{sh} is shown by

$$I_{sh} = \frac{V_s + R_s I_s}{R_{sh}}$$

(30)

where the cell junction temperature T_j is expressed by

$$T_j = T_a + \frac{E_s}{800} (T_{NCOT} - 20)$$

(31)

where P_1, P_2, and P_3 are empirical constants. E_0 and T_0 are reference solar radiation and reference junction temperature, respectively. The P_4 is an empirical constant, a_f is the ideality factor of the PV array, E_g is the gap energy voltage of silicon, κ is Boltzmann's constant, V_s is the terminal voltage, N_s and R_s are the number of cells in series and series resistance, and R_{sh} is the parallel resistance. All the above parameter values for the PV system are shown in Table 2. Figure 6 shows that the maximum power output of the PV system exists at different T_a and E_s.

Table 2: PV Parameters

Symbol	Value
T_0	298 K
T_{NOTC}	316 K
E_0	1000 W m^{-2}
κ	1.3854×10^{-23} J K^{-1}
E_g	1.12 eV
p_1	2.96 Am^{2}W^{-1}
p_2	-8.6×10^{-4} m^{2}W^{-1}
p_3	0.0037 K^{-1}
p_4	1272.3 AK^{-3}
R_s	1.29 Ω
R_{sh}	154.1 Ω

(a)

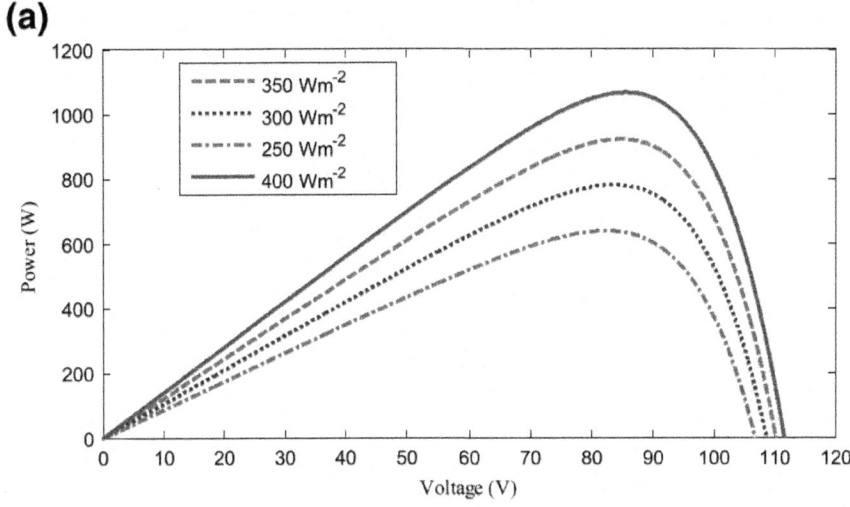

(b)

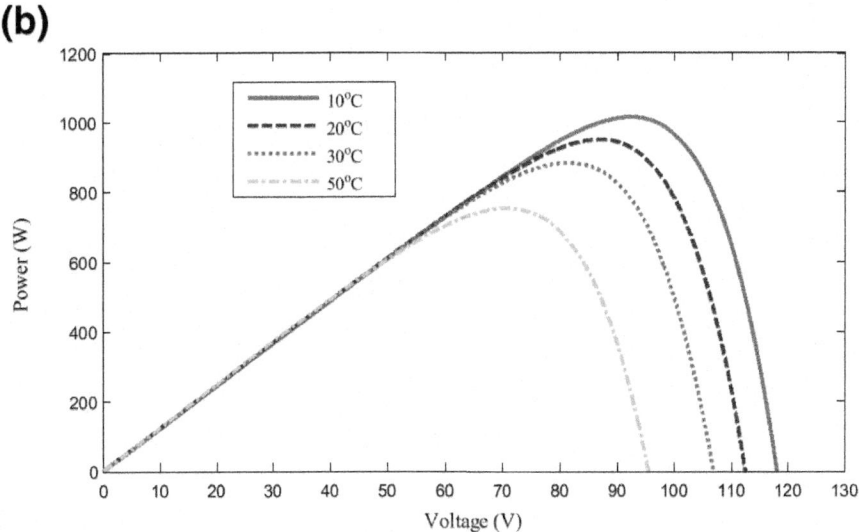

Figure 6: PV power versus voltage (a) at different E_s while T_j =25°C: (b) at different T_a while E_s =350 Wm^{-2}.

Auxiliary Devices

The auxiliary devices shown in Figure 1 include a lithium-ion battery which is considered as a high-capacity energy storage to compensate the power gap and share the load demand and several he boost DC/DC converters are used to boost the DC voltage of the units of PV, PEMFC and battery. The dynamic responses of battery are faster than other units in the hybrid power system, in our approach the modeling of battery is omitted and the well-defined charging/discharging mechanism is considered [16, 17]. The DC/AC pulse-width modulation (PWM) is used to meet the specification of the household electricity, where the dynamics with regard to AC voltage and current waveform are shown in Figure 7 using the SimPowerSystems™ toolbox in Matlab.

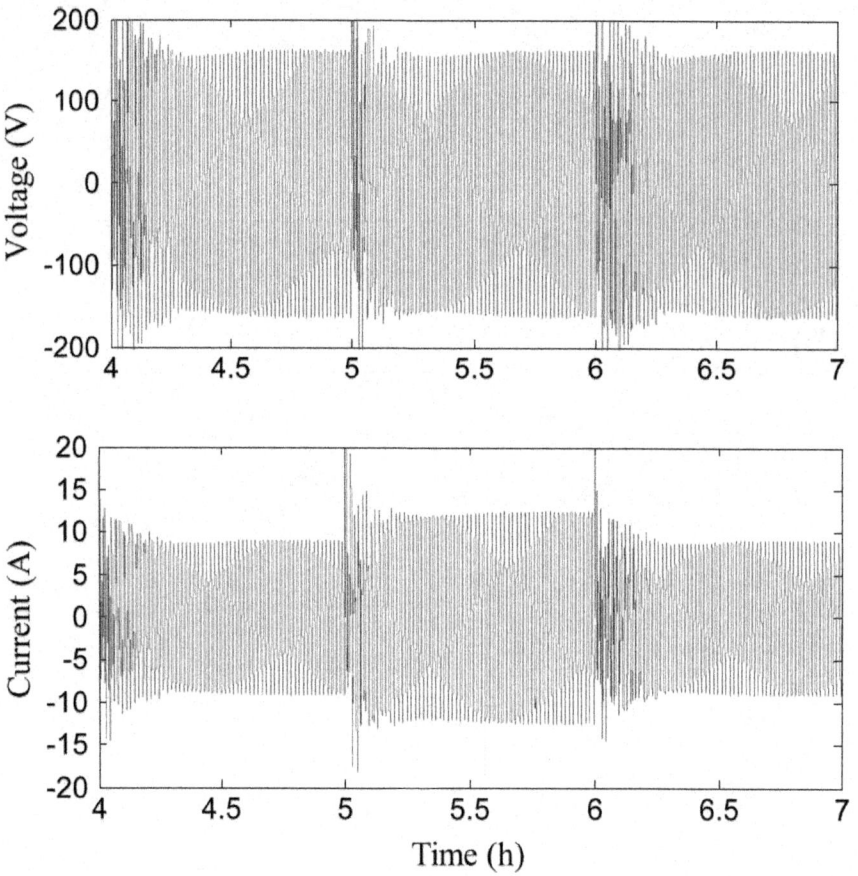

Figure 7: DC/AC PWM inverter connected to DC/DC converter.

Scenario Analysis

For the proposed HPG system, the solar irradiance, water and ethanol are denoted as energy sources. Assumed that the weather variation is inevitable and the sources of water and ethanol are reliable, the HPG system is considered as the optimal power generator since the optimized EFC power generator according to optimal operating conditions in Figures 4 and 5 is specified and the maximum PV power in Figure 6 is achieved by using the maximum power point tracking (MPPT) controller. Moreover, scenario analysis of the daily operation of the HPG system is addressed as follows.

1. Hybrid power dispatch: Figure 8(a) shows that the PV and PEMFC systems are dispatched to precisely meet the daily household load demand [3]. In our design, the daily PV power distribution shown in Figure 8(b) is fixed according to prescribed patterns of solar irradiance. A scenario shows that both PV and PEMFC are integrated to instantly cope with the peak power. Especially, the EFC power should dominate the main electricity supply during the day and night.

2. Power gap compensation: If the EtOH-to-H_2 processor cannot produce pure hydrogen immediately due to slow reactions, then the EFC power decay at this moment. It is caused by the fuel processing delay such that the power demand gap (desired P_{fc}-actual P_{fc}) shown in Figure 8(b) appears. Assumed that the rapid response of Li-ion battery can instantly compensate the power demand gap and the battery charging from PV and EFC units is always larger than the battery discharging, Figure 9(a) and 9(b) show that the HPG system can precisely meet the daily load demand. In this perfect scenario, the hybrid power dispatching shown in Figure 9(a) is confirmed where the battery contribution (green area) aims to compensate the power demand gap during each time period. Figure 9(b) shows that the battery is recharged by the excess power from PV and PEMFC during day and night.

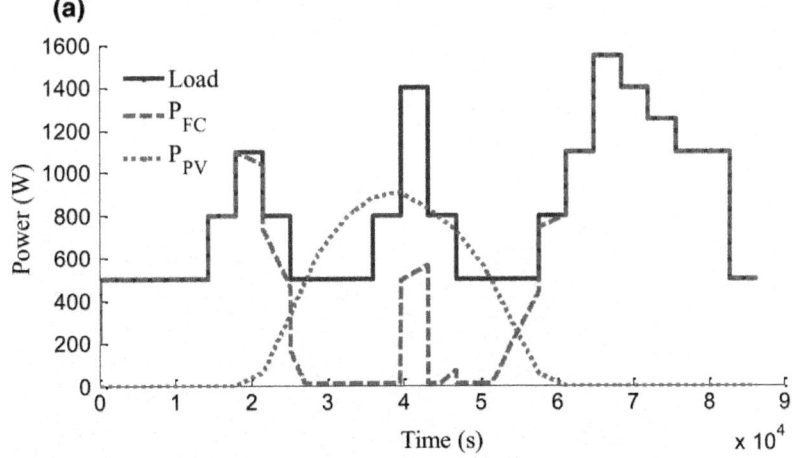

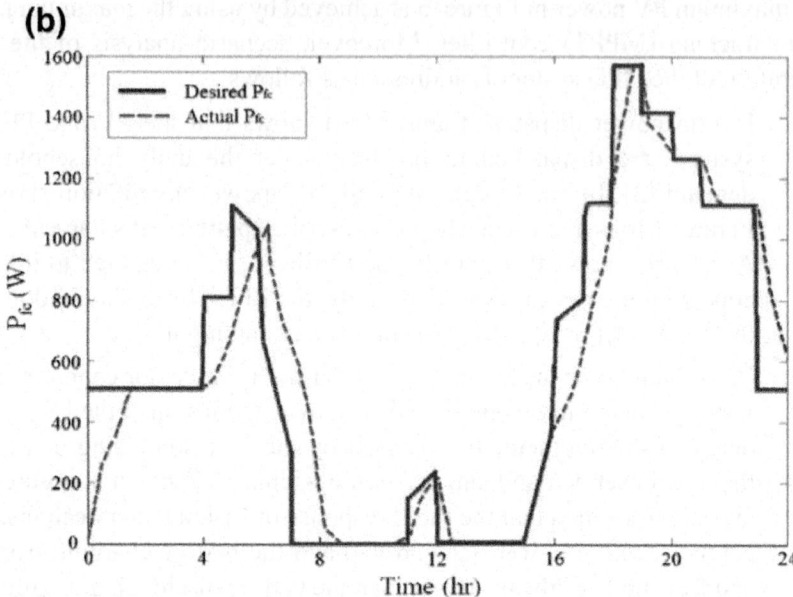

Figure 8: The daily operation of HPG system without use of battery: (a) Power profiles from P_{FC} and P_{PV}; (b) Responses of EFC power generator.

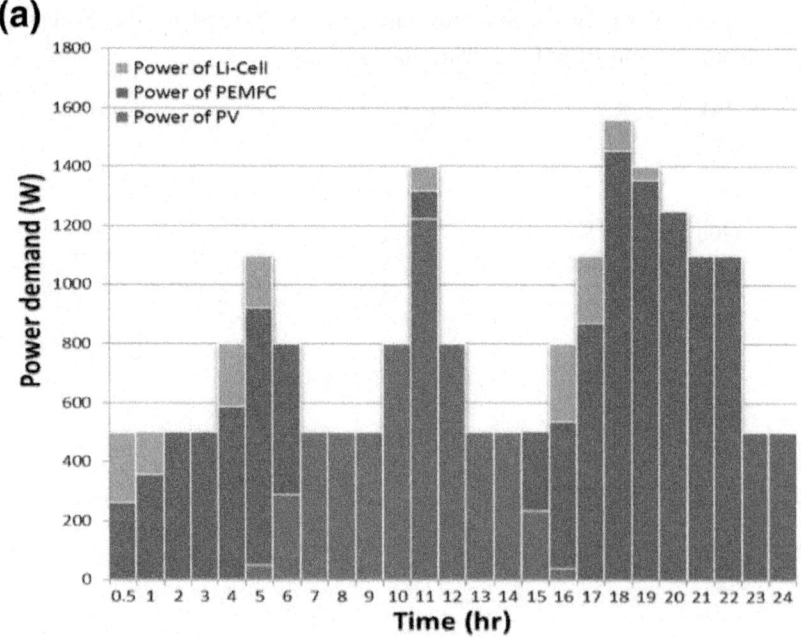

(b)

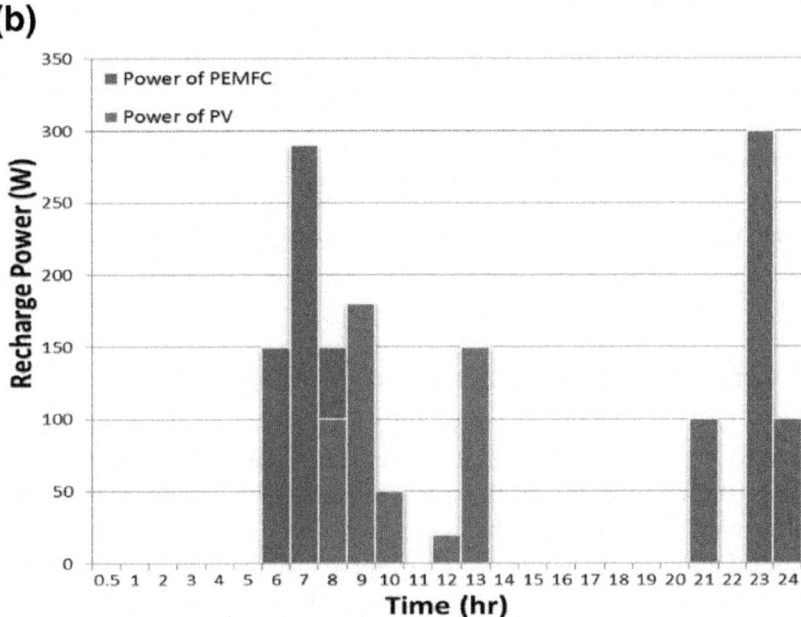

Figure 9: HPG system according to daily power demand: (a) Hybrid power dispatch; (b) Recharge power for battery.

Regarding hybrid power dispatching problem, the PV unit (blue bars) provides 34.5% total power, the battery (green bars) is 12.4% and the EFC (red bars) is about 53.1%. Notably, the battery is charged from the EFC is about 56.1% and 43.9% from the PV. It implies that the PV power is used to save 39.9% fuel consumption.

System Efficiency

The efficiency of the power generation system is usually affected by the electric power, mechanical energy and heat. To ignore the effects of mechanical energy and heat, the efficiency of the HPG system is investigated as follows.

Photovoltaic Efficiency

According to the formulation of solar cell efficiency limits [18], the PV efficiency is described as

$$\eta_{\max} = \frac{P_{\max}}{E_s A_s} \tag{32}$$

where $\eta_{\max}$ is the maximum efficiency of the PV system, $P_{\max}$ is the maximum PV power, and A_s is the surface area of the solar panel.

EFC Efficiency

Referring the experimental data for ethanol reforming processes [19, 20], the energy efficiency of EtOH-to-H$_2$ processor is defined as

$$\eta_{ESR} = \frac{\dot{m}_{H_2,in}HHV_{H_2}}{\dot{m}_{EtOH,in}HHV_{EtOH} + \dot{m}_{H_2O,in}HHV_{H_2O}} \tag{33}$$

and the energy efficiency of fuel cell is described as

$$\eta_{fc} = \frac{P_{fc}}{\dot{m}_{H_2,in}HHV_{H_2}} \tag{34}$$

where HHV$_i$ represent the high heating value of species i. $\dot{m}_{EtOH,in}$ and $\dot{m}_{H_2O,in}$ represent the inlet flow rate of ethanol and water, respectively. Moreover, the efficiency of the EFC is expressed by $\eta_{efc} = \eta_{esr}\eta_{fc}$. Figure 10 shows that the energy efficiency of the EFC unit is increasing if the load demand increases, but the energy efficiency of the FC unit without the use of EtOH-to-H$_2$ processor decreases if the load demand is larger than 5 kW. It is verified that the EFC design can ensure higher energy efficiency than the conventional hydrogen fueled fuel cell system.

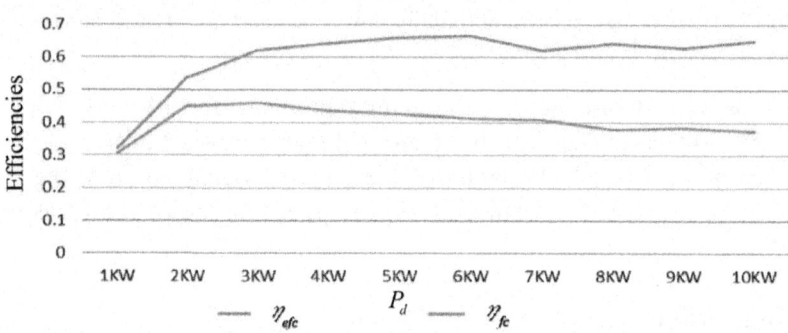

Figure 10: Energy efficiencies for EFC and PEMFC.

According to above efficiency of each unit, the (overall) system efficiency of HPG system is expressed as

$$\eta_{overall} = \frac{(P_{PV} + P_{fc} + P_{battery})\eta_{converter}\eta_{inverter}}{P_{PV}/\eta_{max} + P_{fc}/\eta_{efc} + P_{battery}/\eta_{battery}} \tag{35}$$

where $\eta_{battery}$, $\eta_{converter}$ and $\eta_{inverter}$ are efficiencies of Li-ion battery, DC/DC converter and DC/AC inverter, respectively. The battery power $P_{battery}$ is evaluated by $P_d - P_{efc} - P_{PV}$ where P_d is set as the load. Referring the initial efficiency of Li-ion battery [21], we assume $\eta_{battery}$ for a new battery. Referring to references [22, 23], we assume $\eta_{converter} = 0.9$ and $\eta_{inverter} = 0.98$. Referring the hybrid power dispatching in Figure 9(a), the hybrid system efficiency by Eq. (35) is shown in Figure 11. Based on the hybrid power system using parallel and series connections, the system efficiency is around 11% during the day, but it reaches over 60% during the night. Obviously, the PV system induces lower system efficiency and it can be significantly improved by using the integrated reformer/fuel cell system and battery.

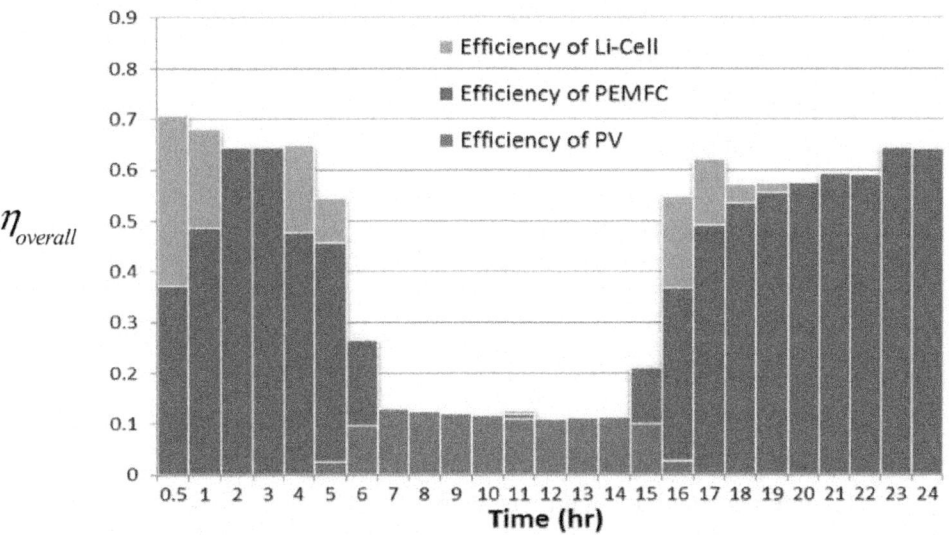

Figure 11: HPG system efficiency.

CONCLUSIONS

An optimized fuel processing unit using ethanol fuel can produce high-purity hydrogen. The simulation shows that the stand-alone EtOH-to-H2 processor not only guarantee higher energy efficiency, but also it can continuously produce hydrogen if the fuel is enough. According to scenarios for the daily operation of the HPG system, the EFC power dominates the power supply during the night, the PV system dominates the power supply during the day and the backup battery aims to instantly compensate the power gap and store the excess power from PV or EFC. According to the hybrid power dispatch, the distribution of the HPG system efficiency is specified.

NOMENCLATURE

a $_f$, ideality factor

A $_{fc}$, effective fuel cell area, cm^2

A $_s$, surface area of the solar panel, m^2

C_{O_2}, oxygen concentration at the cathode/membrane interface, $molm^{-3}$

C_{H_2}, hydrogen concentration at the anode/membrane interface, $molm^{-3}$

C $_{dl}$, Double layer capacitance, F

E $_0$, Reference solar radiation, $W\ m^{-2}$

E $_g$, Gap energy voltage for silicon, eV

E $_s$, solar radiation, Wm^{-2}

i, current density, Acm^{-2}

I $_s$, terminal current of PV cell, A

k $_{an}$, k $_{ca}$, flow constant of anode and cathode, $mol\ s^{-1}\ atm^{-1}$

l $_m$, Membrane thickness, cm

c $_p$, specific heat capacity, $J\ K^{-1}\ kg^{-1}$

P $_i$, partial pressure of i component, atm

P $_{BPR}$, back pressure, atm

P, cell power, W

T $_a$, ambient temperature, K

T $_0$, reference junction temperature, K

T $_{NCOT}$, normal cell operating temperature, K

R $_s$, series resistance, Ω

R $_{sh}$, parallel resistance, Ω

u, superficial velocity, $m\ s^{-1}$

V $_s$, terminal voltage of PV cell, V

Declarations

ACKNOWLEDGMENT

The authors would like to thank the National Science Council of the Republic of China for financially supporting this research under Contract No. NSC 100-2211-E-006-264.

AUTHORS' CONTRIBUTIONS

The energy efficiency of the ethanol-fueled fuel cell (EFC) is higher than the hydrogen-fueled fuel cell system. With the aid of the stand-alone EtOH-to-H2 processor and backup battery. The HPG system efficiency according to hybrid power dispatch is improved. All authors read and approved the final manuscript.

REFERENCES

1. Muselli M, Notton G, Louche A: Design of hybrid-photovoltaic power generator, with optimization of energy management. Sol Energy 1999, 65:143–157. 10.1016/S0038-092X(98)00139-X

2. Elhadidy MA: Performance evaluation of hybrid (wind/solar/diesel) power systems. Renew Energy 2002, 26:401–413. 10.1016/S0960-1481(01)00139-2

3. Hwang JJ, Lai LK, Wu W, Chang WE: Dynamic modeling of a photovoltaic hydrogen fuel cell hybrid system. Int J Hydrogen Energy 2009, 34:9531–9542. 10.1016/j.ijhydene.2009.09.100

4. Uzunoglu M, Onar OC, Alam MS: Modeling, control and simulation of a PV/FC/UC based hybrid power generation system for stand-alone applications. Renew Energy 2009, 34:509–520. 10.1016/j.renene.2008.06.009

5. Zervas PL, Sarimveis H, Palyvos JA, Markatos NCG: Model-based optimal control of a hybrid power generation system consisting of photovoltaic arrays and fuel cells. J Power Sources 2008, 181:327–338. 10.1016/j.jpowsour.2007.11.067

6. Yilanci A, Dincer I, Ozturk HK: A review on solar-hydrogen/fuel cell hybrid energy systems for stationary applications. Prog Energy Combust Sci 2009, 35:231–244. 10.1016/j.pecs.2008.07.004

7. Gupta RB: Hydrogen Fuel: Production, Transport, and Storage. USA: CPC press; 2009.

8. Shekhawat D, Spivey JJ, Berry DA: Fuel Cells: Technologies for Fuel Processing. UK: Elsevier; 2007.

9. Wu W, Tungpanututh C, Yang HT: A conceptual design of a stand-alone hydrogen production system with low carbon dioxide emissions. Int J Hydrogen Energy 2012, 37:10145–10155. 10.1016/j.ijhydene.2012.04.002

10. Garcia VM, Lopez E, Serra M, Llorca J: Dynamic modeling of a three-stage low-temperature ethanol reformer for fuel cell application. J Power Sources 2009, 192:208–215. 10.1016/j.jpowsour.2008.12.055

11. Degliuomini LN, Biset S, Luppi P, Basualdo MS: A rigorous computational model for hydrogen production from bio-ethanol to feed

a fuel cell stack. Int J Hydrogen Energy 2012, 37:3108–3129. 10.1016/j. ijhydene.2011.10.069

12. Uriza I, Arzamendi G, Lopez E, Llorca J, Gandia LM: Computational fluid dynamics simulation of ethanol steam reforming in catalytic wall microchannel. Chem Eng J 2011, 167:603–609. 10.1016/j.cej.2010.07.070

13. Choi Y, Stenger HG: Water gas shift reaction kinetics and reactor modeling for fuel cell grade hydrogen. J Power Sources 2003,124:432–439. 10.1016/S0378-7753(03)00614-1

14. Khan MJ, Iqbal MT: Analysis of a small wind-hydrogen stand-alone hybrid energy system. Appl Energy 2009, 86:2429–2442. 10.1016/j. apenergy.2008.10.024

15. Wu W, Xu JP, Hwang JJ: Multi-loop nonlinear predictive control scheme for a simplistic hybrid energy system. Int J Hydrogen Energy 2009, 34:3953–3964. 10.1016/j.ijhydene.2009.02.060

16. Knauff M, McLaughlin J, Dafis C ASNE Intelligent Ships Symposium, Philadelphia, PA, USA. Simulink model of a lithium-ion battery for the hybrid power system test-bed 2007.

17. Erdinc O, Vural B, Uzunoglu M 2nd International Conference on Clean Electrical Power, Capri, Italy. A dynamic lithium-ion battery model considering the effects of temperature and capacity fading 2009.

18. Nelson J: The Physics of Solar Cell. UK: Imperial College Press; 2003.

19. Deluga GA, Salge JR, Schmidt LD, Verykios XE: Renewable hydrogen from ethanol by autothermal reforming. Science 2004,303:933–937.

20. Hung CC, Chen SL, Liao YK, Chen CH, Wang JH: Oxidative steam reforming of ethanol for hydrogen production on M/Al2O3. Int J Hydrogen Energy 2012, 37:4955–4966. 10.1016/j.ijhydene.2011.12.060

21. Shen J, Tang Y, Liang Y, Tan X: Relationship between initial efficiency and structure parameters of carbon anode material for Li-ion battery. J Cent South Univ Technol 2008, 15:484–487. 10.1007/s11771-008-0091-y

22. Ma D, Ki WH, Tsui CY, Mok PKT Proceeding of the ASP-DAC, Asia and South Pacific. A single inductor dual-output integrated dc/dc boost converter for variable voltage scheduling 2001.

23. Anis RA: Stepped sine wave DC/AC inverter I. Theoretical analysis. Energy Mater Sol Cells 1992, 28:123–130. 10.1016/0927-0248(92)90004-9

Chapter 10

SYSTEMATIC FRAMEWORK FOR MULTIOBJECTIVE OPTIMIZATION IN CHEMICAL PROCESS PLANT DESIGN

Ramzan Naveed[1], Zeeshan Nawaz[2], Werner Witt[3] and Shahid Naveed[1]

[1]Department of Chemical Engineering, University of Engineering and Technology, Lahore, Pakistan

[2]Chemical Technology Development, STCR, Saudi Basic Industries Corporation (SABIC), Kingdom of Saudi Arabia

[3]Lehrstuhl Anlagen und Sicherheitstechnik, Brandenburgicshe Technische Universität, Cottbus, Germany

INTRODUCTION

For solving multiobjective decision making problems, a systematic and effective procedure is required. As far as the process or control system has to be modified process simulators like Aspen Plus™, Aspen Dynamics are widely used. But these simulators are not designed for investigation of other objectives as environment and safety. Due to complex and conflicting nature of multiobjective decision making an integrated optimization tool should be of value. In this chapter a systematic methodology based on independent modules and its different stages to deal this problem is presented in detail.

PROPOSED METHODOLOGY

The methodology is built around several standard independent techniques. These techniques have been suitably modified/adopted and woven together in an integrated plate form. The main aim is to standardize the screening and selection of decisions during design/modification of chemical process plant and optimizing the process variables in order to generate a process with improved economics along with satisfaction of environmental and safety constraints. The methodology (see Figure 1) consists of four layers/stages:

- Generation of alternatives and problem definition;
- Analysis of alternatives i.e. generation of relevant data for comparison of Environmental, economic and safety objectives

- Multiobjective decision analysis/ optimization
- Design evaluation stage i.e. decision making from the pareto-surface of non-inferior solution or ranking of alternatives

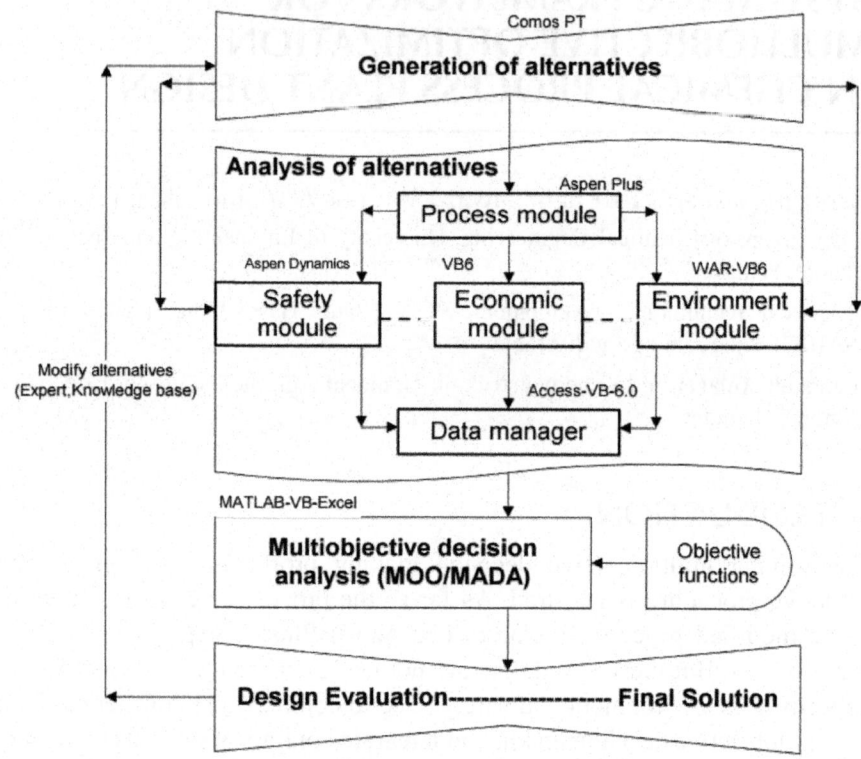

Figure. 1. Simplified block diagram of proposed methodology.

Stage I: Generation of Alternatives and Problem Definition

The first layer composed of following tasks:

Definition of the scope of the study,

Statement of key assumptions and the performance targets such as quality etc.,

Degree of freedom analysis,

Identification of the key design, control, and manipulated variables,

Definition of the system boundary,

Identification of constraints,

Choice of functional unit for all calculations,

Collection of relevant information about process and chemicals to be handled,

Generation of different alternatives either based on suggestion from independent departments or using the individual objective modules from stage II.

Seader et al. (1999)[6] has described rules for selection of process variables in the book

"Process design principles-synthesis, analysis and evaluation". The data and information about the process and chemicals involved such as thermodynamic and kinetic data can be found from journal articles, patents or handbooks. Current chemical prices can be obtained from market reports if not available in main plant documentation or company central data base. In addition to these sources, some data related to quantification of environmental impacts and material safety data sheets of chemical are also collected from commercial data bases so that an impact assessment and safety analysis can be performed in subsequent design steps. Commercial computer aided tool like ComosPT can be used for plant documentation and to support stage-I of proposed methodology.

Stage II: Analysis of Alternatives

This stage is composed of independent modules used to generate relevant information for evaluation of economic, safety and environmental performance objectives. These modules are:

- Process module
- Safety module [1,3]
- Economic module
- Environment module [2]
- and a data manager for managing the relevant information generated from these modules.

Process Module

In the process module, an operation model of the process system has to be developed for evaluating alternatives. The configured simulation model has to be able to reproduce the selected results to an accepted degree of accuracy. This simulation model can be used for design and operation, revamping and debottlenecking of the process under study[7]. Three major integrated simulation systems widely used in the firms and companies for this purpose are Aspen technology (Aspen Plus, Aspen dynamics etc), Hyprotech (Hysys process, Hysys plant etc) and Simulation Sciences (Pro/II etc.). Aspen Plus™ 12.1 is used in this work for development of simulation model and linked in a visual basic platform for integration with safety, economic and environment modules. The most important results available from the process simulation model are material and energy balance information for both streams and units, rating performance of units and tables and graphs of physical properties. A brief description of Aspen Plus™ 12.1 and steps involved in development of the process simulation model is described here below.

Aspen Plus™

Aspen Plus™ supports both sequential modular and equation oriented computation strategy and allows the user to build and run a steady-state simulation model for a chemical process. It provides a flexible and productive engineering environment designed to maximize the results of engineering efforts, such as user interface mode manager, quick property analysis, rigorous and robust flowsheet modelling, interactive architecture, powerful model analysis tools and analysis and communication of results. Therefore, it lets the user to focus his/her energies on solving the engineering problems, not on how to use the software. It is not only good for process simulation but also allows to perform a wide range of other tasks such as estimating and regressing physical properties, generating custom graphical and tabular output results, sensitivity analysis, data-fitting plant data to simulation models, costing the plant, optimizing the process, and interfacing results to spreadsheets.

The development of a simulation model for a chemical process using Aspen PlusTM 12.1 involves the following steps (see details in table 1):

1. Define the process flowsheet configuration by specifying
   a. Unit operations
   b. Process streams flowing between the units
   c. Unit operation models to describe each unit operation
2. Specify the chemical components,
3. Choose a thermodynamic model to represent the physical properties of the components and mixtures in the process,
4. Specify the component flow rates and thermodynamic conditions (i.e. temperature, pressure, or phase condition) of the feed streams, 5. Specify the operating conditions for the unit operations,

Table 1. Developmental process for an Aspen PlusTM simulation model

Step	Used to
Defining the flowsheet	Break down the desired process into its parts: feed streams, unit operations, and product streams
Specifying stream properties and units	Calculate the temperature, pressure, vapor fraction, molecular weight, enthalpy, entropy and density for the simulation streams
Entering components	From a databank that is full of common components
Estimating property parameters	Property Constant Estimate System (PCES) can estimate many of the property parameters required by physical property models
Specifying streams	Streams connect unit operation blocks in a flowsheet and carry material and energy flows from one block to another. For all process feed streams, we must specify flowrate, composition, and thermodynamic condition
Unit operation blocks	We choose unit operation models for flowsheet blocks when we define our simulation flowsheet

Safety Module

Safety module is based on combination of conventional standard risk analysis techniques and process disturbance simulation. This module not only generates relevant information related to safety aspects for multiobjective decision analysis but also used for safety/risk analysis and optimization. The purpose of this module is to determine risk from operational disturbances and to develop effective risk reductions. It can be divided into the following steps (Figure 2):

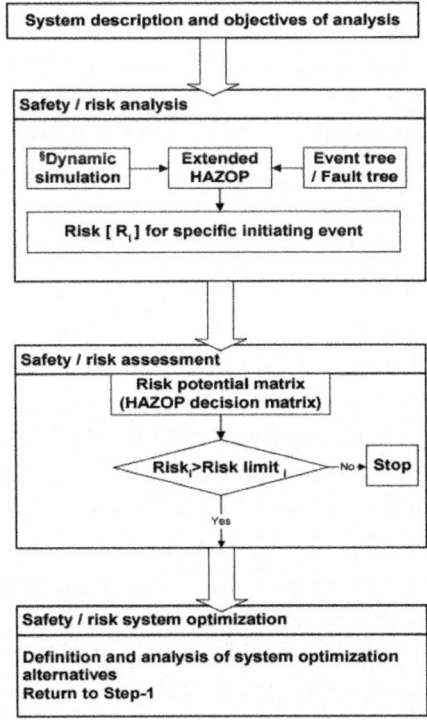

§ Simulation of process related malfunctions

i = { financial risk , environmental risk , human health risk }

Step 1: System description and objectives of analysis (before starting safety and risk analysis)

Step 2: Safety/risk analysis (identification of weak points via Extended HAZOP)

Step 3: Safety/risk assessment (categorization of risk via risk potential matrix (HAZOP decision matrix))

Step 4: Safety/risk system optimization

Figure. 2. Simplified block diagram of safety module

Step 1: (Before Starting Safety/Risk Analysis) - Description of System and Objectives of Analysis

For efficient safety/risk studies, the analyst must have an accurate description of the system to be investigated and a clear objective of the analysis study. Therefore, in this step the purpose, objectives, and scope of the study are clearly defined. The necessary information required for the study such as process flow diagrams, piping and instrumentation diagrams, plant layout schematics, material safety data sheets, equipment data sheets, operating instructions, start up and emergency shutdown procedures, and process limits, etc. is gathered from plant documentation. A team under a trained and experience leader with five to seven people including experts of the design and operation of the subject process may be formulated.

Step 2: Safety/Risk Analysis (Identification of Weak Points Via Extended HAZOP) - Extended HAZOP

Our intention is to identify weak points due to disturbances in operation, which may or may not be hazardous, in order to improve safety, operability, and/or profit at the same time. Extended HAZOP (HAZOP supported by dynamic simulation, event tree and fault tree techniques and HAZOP decision matrix) is used not only for identification of weak points but also for generation and analysis of optimization proposals [8-11]. Extended HAZOP differs from the standard HAZOP approach in following aspects:

- Use of dynamic simulation: In Extended HAZOP, the analysis of the influence of disturbances (failures) on the behavior of the process is based on shortcut or simplified hand calculations or dynamic simulation. Aspen dynamics is used for this purpose.

- Classification of risk related consequences: Each established consequence (hazard) has to be expressed by a consequence class (C). The plant specific scoring (from 0 (lowest) to 8 (highest)) chart is given in Table 2 (a & b) based on principle consequence analysis. For classification of consequences based on principle release estimates, accident consequence analysis techniques (models for calculation of toxic, fire and explosion effects) and plant location data (capital investment, population density etc.) have to be considered.

Illustrative Example 1

Figure 3 shows the plant lay out considered for developing plant specific consequence scoring chart. The area around the plant is open fields (rural condition). As weather conditions changed around the year, so certain assumptions are made to results in worse case conditions for consequence analysis. These include weather conditions and wind speed that result in smallest value of dispersion coefficients. Therefore, stability "F" and wind speed as low as possible (1.5 m/s) is selected. It is assumed that 10 workers are present (working 24 h each day), which are not distributed uniformly, on the land in area (100 m x 100 m) around the column under study. Acetone is selected as representative fluid for consequence analysis.

Acetone vapors released from the vent line at a rate of 1616 kg/h due to loss of cooling medium. It is assumed that released vapors form a cloud for 30 minutes before being ignited and leads to vapor cloud explosion

Table 2. Scoring chart for Consequence Financial consequences [3.4]

Effects	Class	Financial loss (€)	Class related consequences: examples
Function impairment	*	< 10	: Product quality lowering (brief)
	1	$10^1 - 10^2$	: Product quality lowering
	2	$10^2 - 10^3$	: Product quality lowering (long term)
Functional Loss	3	$10^3 - 10^4$	: Production disturbance (brief) Soil contamination Safe dispersion of material release from vent line
	4	$10^4 - 10^5$	: Production disturbance Material release from the piping Pump damage (pressure impacts)
	5	$10^5 - 10^6$	: Production disturbance (long term) Jet fire as result of release of material from vent line Pool fire (from pump leakage)
Safety and Environmental pollution	6	$10^6 - 10^7$	Fireballs due to catastrophic rupture of vapour product line
	7	$10^7 - 10^8$	Vapour cloud explosion (ignoring domino effect)
	8	$>10^8$	Vapour cloud explosion along with domino effect

Effects	Class	Community	Class related consequences: examples
Function impairment	*	No effect on people	: Product quality lowering (brief)
	1	Nuisance effect	: Product quality lowering
	2	Minor irritation effect to people & local news	: Product quality lowering (long term)
Functional Loss	3	Moderate irritation effect to people and non compliance to laws, local news	: Production disturbance (brief) Soil contamination Safe dispersion of material release from vent line
	4	Moderate irritation effects to people & environment, single injuries and regional news	: Production disturbance Material release from the piping Pump damage (pressure impacts)
	5	Significant effects to people and environment, > 1 injuries & regional news	: Production disturbance (long term) Jet fire as result of release of material from vent line Pool fire (from pump leakage)
Safety and Environmental pollution	6	Major effects to people and environement, multiple injuries, fatality likely, regional news	Fireballs due to catastrophic rupture of pipe or condenser (vapour product line)
	7	Severe effects to people and environment, fatality, regional news	Vapour cloud explosion (ignoring domino effect)
	8	Multiple fatalities and process shutdown certain, international news	Vapour cloud explosion along with domino effect

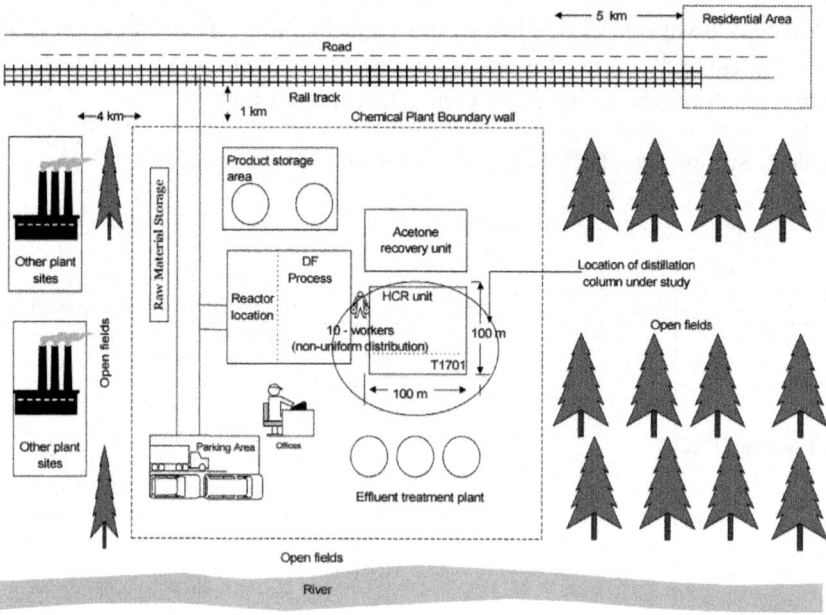

Figure. 3. Plant lay out for establishing consequence score chart.

The physical effects of this scenario or event is calculated as:

Weight of fuel in the cloud= M = 1616 / 2 = 808 kg

Then amount of TNT equivalent to the amount of this flammable material is

$$M_{TNT} = \alpha \cdot \frac{M \cdot H_c}{H_{TNT}}$$

Where a = explosion efficiency ~ 0.05 (Cameron 2005) Hc = heat of combustion of fuel ~ 3.03 x 104 kJ/kg for acetone H_{TNT} = TNT blast energy ~ 5420 kJ/kg; so 225.25 M_{TNT} = kg

Then, using relation $Z=R/M_{TNT}^{1/3}$ and figure 4, scaled distance and overpressure is estimated. Table 3 presents the results obtained.

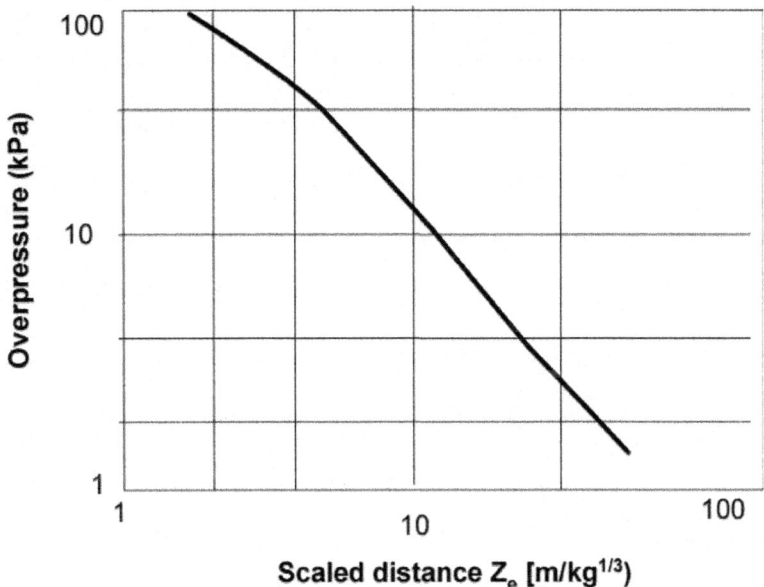

Figure. 4. Overpressure versus scaled distance for TNT explosions on flat surfaces (Tweeddale 2003, p. 115).

Table 3. Results of physical effects of vapour cloud explosion

Distance , R M	Scaled distance, Z m / kg1/3	Overpressure, Δp kPa
10	1.64	90
20	3.28	40
50	8.21	20
100	16.40	7

It is estimated that severe structural damage and 15 % chance of fatality outdoors or 50 % chance indoor will be experienced out to 20 m and almost complete destruction of all ordinary structures and 100 % chance of fatality indoors to 10 m distance.[8] (see Cameron 2005, p. 268).

iii.   Classification of frequencies of risk related consequences: The frequency of occurring for each possible consequence (hazard) has to be expressed by a frequency class, called (F) according to the scoring chart for frequency (Table 3.4). Definition of frequency class may be supported by Event Tree and/or Fault tree analysis techniques or Layer of protection analysis (LOPA) or historical databases.

For establishing frequency class: Estimation / calculation of frequency of vapor cloud explosion and fatality of person because of release of material due to catastrophic rupture of distillation column.

Frequency of catastrophic rupture of column = 10-6 (Taken from table 4)

Table 4. Scoring chart for frequency [3.4]

Class	Frequency 1/y	Comprehension	Examples based on general data bases
		Frequency of occurring incident	
9	<10⁻⁸	Very very small	Catastrophic rupture or leakage of pipe of diameter > 150 mm
8	10⁻⁸ - 10⁻⁷	Very small	Catastrophic rupture of pipe of diameter ≤ 50 mm
7	10⁻⁷ - 10⁻⁶	Small	Catastrophic rupture of fractionating system (excluding piping), storage tank rupture
6	10⁻⁶ - 10⁻⁵	Less small	Pipe residual failure, 100 m full breach, Double wall tank leakage
5	10⁻⁵ - 10⁻⁴	Moderate	Process vessel leakage of ≥ 1 mm diameter
4	10⁻⁴ - 10⁻³	Less moderate	Pump leakage , Heat exchanger leakage
3	10⁻³ - 10⁻²	Less high	Safety valve open spuriously, Large external fire
2	10⁻² - 10⁻¹	High	Cooling water failure, BPCS instrument loop failure
1	10⁻¹ - 10⁰	Very high	Operator failure, Regulator failure , Solenoid valve failure
*	>10⁰	Very very high	Power failure in developing countries, Operators failure under high stress

Probability of ignition of released material = 0.10 (CCPs 2000, Borysiewich 2004)

Probability of VCE if released material ignited = 0.01 (CCPs 2000, Borysiewich 2004)

Probability of fatality of a person exposed to overpressure of 40 kPa due to VCE = 0.20 (Tweeddale 2003, p. 117 Figure 5-14)

Then, frequency of vapour cloud explosion = $10^{-6} \cdot 0.10 \cdot 0.01 = 10^{-9}$

So frequency class for this scenario = 9

Frequency of fatality of a person exposed to VCE = $10^{-9} \cdot 0.20 = 2.10^{-10}$

So frequency class for this scenario = 9

iv.   Way of documenting the HAZOP results

The Extended HAZOP methodology worksheet for documenting the HAZOP team results is shown in Figure 5. Below consequence the physical effects and risk has to be documented first and next risk has to be classified using score charts (Table 3.2) related to financial, environment and health related consequences. The worst score of each risk has to be documented. For each risk related consequence, frequency class has also to be established.

Consequence, €		<10	$10^1 -$ 10^2	$10^2 -$ 10^3	$10^3 -$ 10^4	$10^4 -$ 10^5	$10^5 -$ 10^6	$10^6 -$ 10^7	$10^7 -$ 10^8	>10^8
Frequency C 1/y F		*	1	2	3	4	5	6	7	8
>10^0	*									
$10^{-1} - 10^0$	1									
$10^{-2} - 10^{-1}$	2									
$10^{-3} - 10^{-2}$	3									
$10^{-4} - 10^{-3}$	4									
$10^{-5} - 10^{-4}$	5									
$10^{-6} - 10^{-5}$	6									
$10^{-7} - 10^{-6}$	7									
$10^{-8} - 10^{-7}$	8									
<10^{-8}	9									
		Immediate action needed before further operation								
		Action at next occasion after qualification of analysis for improving system								
		Optional								
		No further action needed								

Figure. 5. Risk potential matrix (Extended HAZOP decision matrix).

Illustrative Example 2

Release of material to atmosphere from vent line or vapour line may disperse safely or has toxic effects or can lead to several outcomes such as flash fire, vapour cloud explosion and fire balls. So documenting consequence class in HAZOP work sheet, the score '8' of the most severe consequence will be documented.

Plant/P&ID : Equipment : Volume :		Process: Function:				Document: Page : Date :		
No	Guide word / Process Parameter	Detection/ Safeguards	Possible causes	Conse-quences	FC	Recommended Actions	FC	Resp./ Ref.
				Physical effects: Risk related:				

1- Short cut calculations 2-Dynamic simulation 3-deterministic models 4-Event tree 5- Fault tree 6- Historic data base

Step 3: Safety/Risk Assessment - Risk Potential Matrix (HAZOP Decision Matrix)

Figure 6 shows the risk potential matrix (HAZOP decision matrix) used for order of magnitude ranking of events. The rows of the matrix consider frequency class, while the columns show the consequence class. Each cell in the matrix represents a risk category. For the decision process, the matrix is divided into four risk category levels.

Risk level I --- red area --- scenario in this level is intolerable and immediate action (pant or process modification) is needed to reduce that risk category or more detailed quantified analysis has to be carried out in order to find arguments for wrong preliminary decisions.

Risk level II --- grey area --- scenario in this level is tolerable but not acceptable for long period of time so action at next schedule maintenance is needed to reduce that risk category.

Risk level III --- yellow area --- scenario in this level is acceptable and any action to reduce that risk category is optional.

Risk level IV--- green area --- scenario in this level needs no action.

Risk potential matrix (HAZOP decision matrix) may be also used for:

- Documentation of the status of the plant safety
- Selection and development of optimization proposals
- Importance of improvement
- Documentation of improvement achieved

The application of risk potential matrix (HAZOP decision matrix) in the Extended HAZOP is shown in figure 5. Arrows show the transformation of entries from the Extended HAZOP worksheet to the HAZOP decision matrix. The identity number (ID) of each scenario of the Extended HAZOP worksheet is placed in HAZOP decision matrix. Recommended actions for this scenario will be placed from Extended HAZOP sheet to the bottom of HAZOP decision matrix. First HAZOP decision matrix will shows the existing status and second HAZOP decision matrix shows the improved plant status after recommended actions.

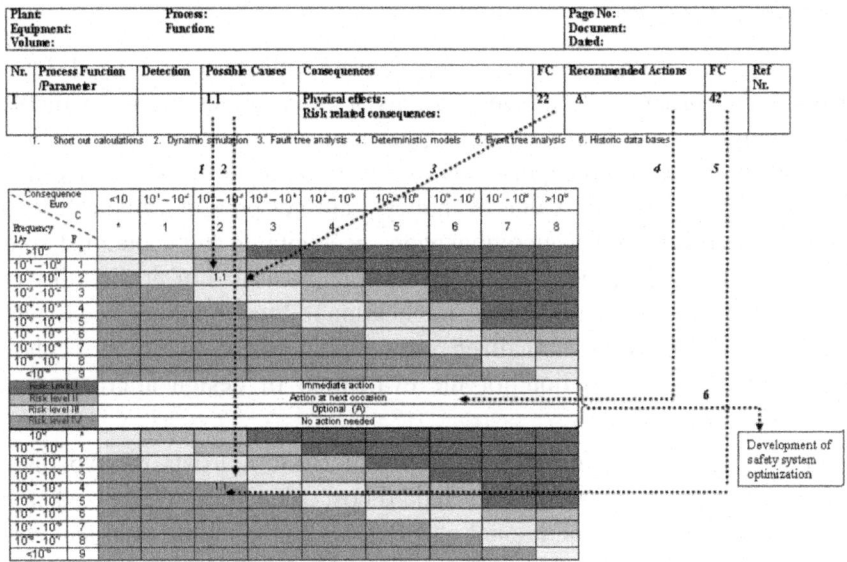

Figure. 6. Application of HAZOP decision matrix in Extended HAZOP.

Similarly all results from Extended HAZOP worksheets are transferred to the HAZOP decision matrix. Keeping in view the risk target and depending on the scenario or recommended actions during the Extended HAZOP discussion, analysis team may reach a safety related modification proposal. Next, if safety/ risk optimization is in focus then weak points/scenarios with similar risk are clustered after analyzing HAZOP decision matrix and safety related optimization proposals are developed.

Step 4: Safety/Risk System Optimization (Development and Analysis of Optimization Proposals)

In this step, safety related optimization proposals are generated and evaluated using dynamic simulation, Event tree analysis and/or Fault tree analysis. The optimization proposals can be developed at two levels:

- Simple optimization proposals e.g. addition of pressure alarm or change of location of sensor within the Extended HAZOP discussion
- Optimization proposals related to severe scenarios by evaluating risk potential matrix (HAZOP decision matrix)

The relevant information such as frequency and damage data will be transferred to economic module for safety related cost calculations and multiobjective decision making (if more than one alternatives developed).

Economic Module

In all stages of design process, economic evaluation is crucial for the evaluation of process alternatives. Various objective functions are available in the literature of chemical engineering economics for economic evaluation of chemical processes. Some quite elegant objective functions, which incorporate the concept of the „time value of money", are net present value (NPV) and discounted cash flow. Business managers, accountants and economists prefer these methods because they are more accurate measures of profitability over an extended time period. However, application of these methods needs certain assumptions[12]. Total annualized cost (TAC) can be used as economic indicators/objective function for the evaluation of design alternatives and economic optimization.

Economic module developed in Visual Basic consists of two distinct sections. First section carries out standard cost calculations (i.e. Fixed capital investment (FC1) and operational cost (OC1)) and compute total annualized cost (TAC1) while second section carries out extended cost calculations i.e. process safety/risk related costs and computes the fixed capital investment related to safety system (FCISS), accident and incident damage related risk cost. Table 3.5 illustrates the difference of cost elements considered in standard practice of cost calculations of chemical process design and in this economic module. Figure 7 shows the simplified block diagram of economic module.

Standard Cost Calculations

Standard cost calculations involves fixed capital investment (FCI1) and operational cost (OC1). Fixed capital investment (FCI1) includes the cost of design and other engineering and construction supervision, all items of equipment and their installation, all piping, instrumentation and control systems, buildings and structures, and auxiliary facilities such as utilities, land and civil engineering work.

Table 5. Elements of economic module and difference from standard cost calculations

Standard cost calculations	Economic module used in this work

• FCI1 = Fixed capital investment using either cost equations that have been derived by Ulrich or correlations developed by Guthrie depending on users choice
• OC1= Operating cost (including both direct (e.g. raw material, utilities etc.) and indirect costs (e.g. taxes, overhead cost etc)
• TAC1 = total annualized cost
   = d·(FCI1)+OC1
normally d is taken 0.15-25 but can also be computed using depreciation calculation methods
	• FCI1= Fixed capital investment using either cost equations that have been derived by Ulrich or correlations developed by Guthrie depending on users choice
	• FCI2= FCI1 + fixed capital investment related to safety system (FCISS)
	• OC1= Operating cost (including both direct (e.g. raw material, utilities etc.) and indirect costs (e.g. taxes, overhead cost etc)
	• TAC1 = total annualized cost
	   = d·(FCI1)+OC1
	• TAC2 = total annualized cost
	   = d·(FCI2)+OC1
	normally d is taken 0.15-25 but can also be computed using depreciation calculation methods

	Extended cost calculations • RC1 = risk cost 1= Asset risk cost + health risk cost + environmental risk cost • RC2 = risk cost2 = RC1+ production loss risk cost • RC3 = risk cost3 = process interruption cost • TRC = total risk cost = RC2+RC3 • ECC = Extended costs

Several capital cost estimate methods ranging from order of magnitude estimate (ratio estimate) to detailed estimate (contractors estimate) are used for the estimation of installed cost of the process units in the chemical plant. The most commonly used method that provides estimates within 20-30% of actual cost and widely used at design stage involve the usage of cost charts/correlations (Guthrie's article (1969) and book (1974), chapter 5 of Ulrich's 'A guide to chemical engineering process design and economics' (1984), 'Plant design and economics for chemical engineers' by Peters and Timmerhause (1991)) for estimating the purchase cost of major type of process equipment [13-15].

These cost charts / correlations were assembled in the 1960's or earlier and are projected to the date of installation using cost indices or escalation factors such as the chemical engineering plant cost index (published biweekly by chemical engineering magazine),

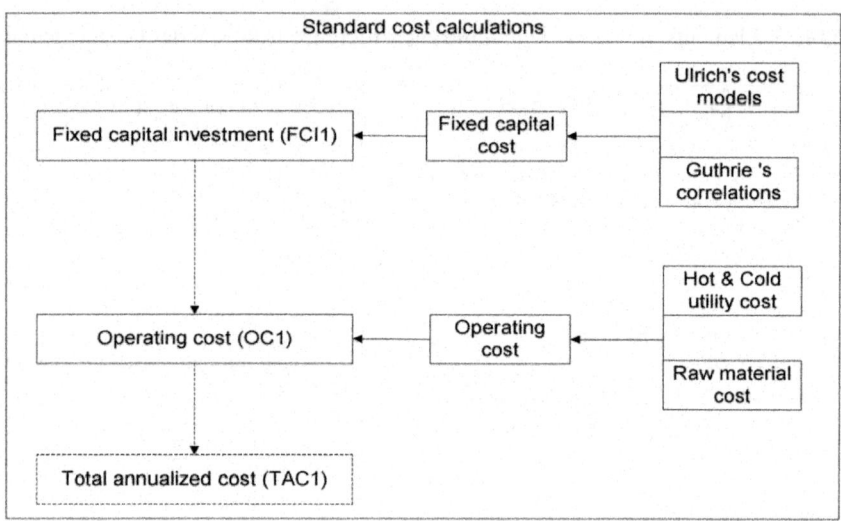

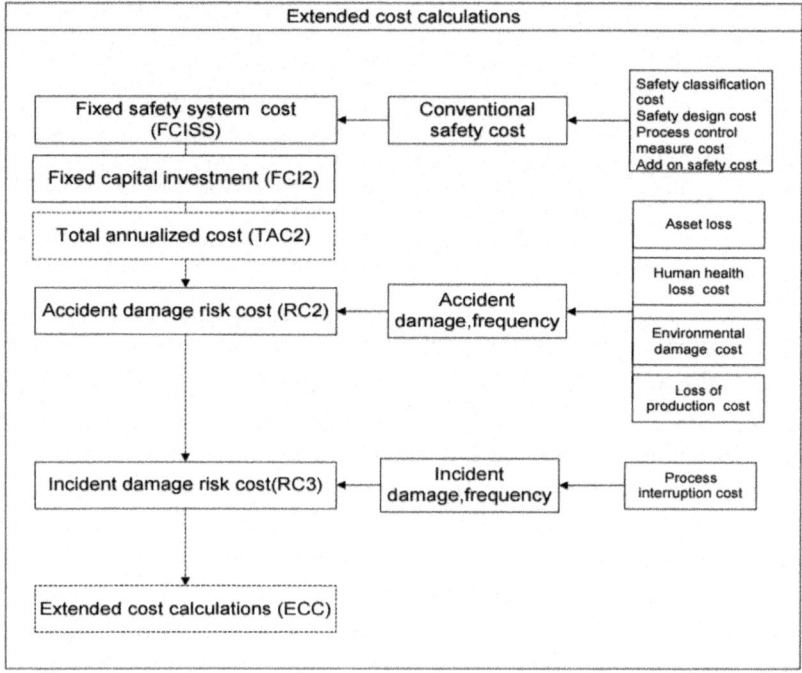

Figure. 7. Simplified block diagram of economic module.

Note: In Extended cost calculations, the costs such as insurance cost, market loss cost, loss of image and prestige cost should also be considered in addition. But in this module these costs are not included.

Marshall and Swift Index (also provided in chemical engineering magazine) and NelsonFabaar Index (from the oil and gas journal). For the comparison of process design alternatives, these study estimates for purchased cost of process units using cost charts or equations based on them are adequate. Given the purchase cost of a process unit, the installed cost is obtained by adding the cost of installation using factored-cost methods. For each piece of equipment Guthrie (1969, 1974) provides factors to estimate the direct cost of labor, as well as, indirect costs involved in the installation procedure. The cost elements that are included in the estimation of fixed capital investment are shown in figure 8.

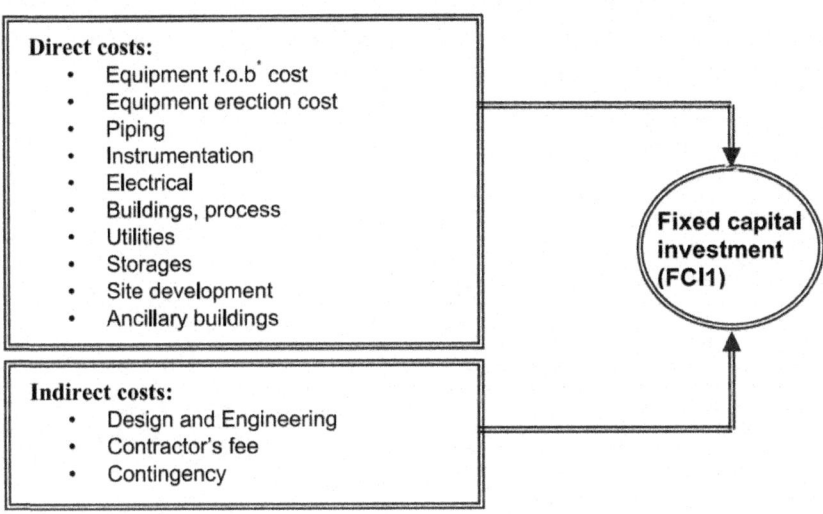

Figure. 8. Typical cost elements for fixed capital investment. (f.o.b ---- freight on board cost i.e. equipment purchase cost plus transport cost).

The operating cost (OC1) of a chemical plant is divided into two groups:

- Fixed operating cost
- Variable operating cost

The elements in fixed operating cost includes maintenance cost, operating labor cost, laboratory cost, supervision cost, plant overheads, capital charges, taxes, insurance, licence fees and royalty payments while the variable operating cost consists of raw material costs, miscellaneous operating material costs, utilities (services) and shipping and packaging. However this division of operating cost is somewhat arbitrary and depends on the accounting practice of a particular organization. The typical cost elements included in operating cost "OC1" are shown in figure 9.

However, from the existing process optimization point of view, energy cost and raw material costs are more important and often considered.

Economic module developed in this thesis using Visual basic computes fixed capital expenditure using either cost equations that have been derived by Ulrich or correlations developed by Guthrie depending on users choice, The significant operating cost for process optimization i.e. energy consumption cost (heating and cooling utilities cost) can also be calculated by using this module. Once the FCI1 and OC1 are calculated, then total annualized cost is obtained using the following equation:

Total annualized cost $(TAC1)$ d $(FCI1)$ $OC1 = \times + (1)$

Here d is depreciation or capital recovery factor and normally taken between 0.15-0.25 but can also be computed using depreciation calculation methods e.g. double declining balance method.

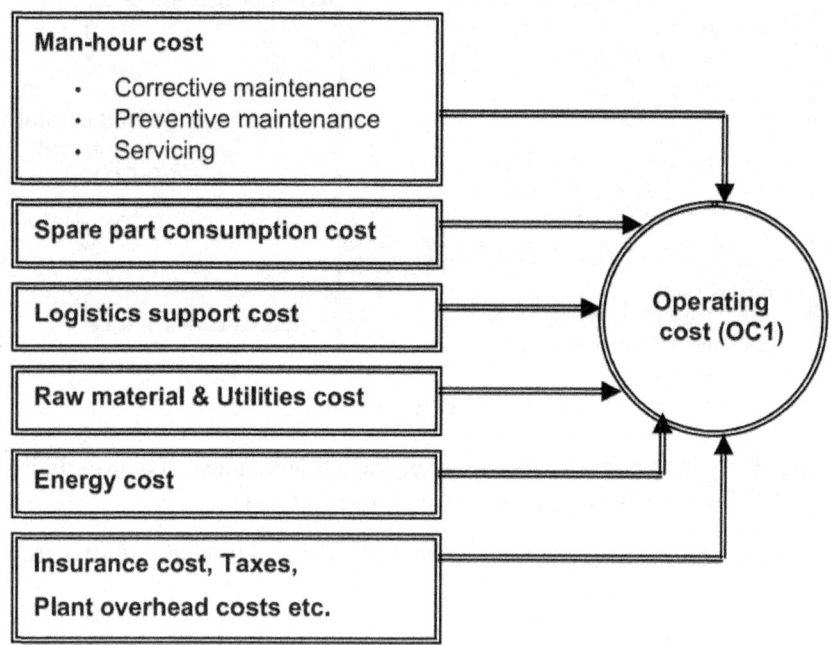

Figure. 9. Typical cost elements for operating cost (OC1).

Extended Cost Calculations

The second section of the economic module (see Figure 7) carries out Extended cost calculations, which considers the fixed capital investment related to safety system, and risk cost due to accident and incident damage.

i.   Fixed capital investment related to Safety system: The fixed investment related to safety system is calculated by the following equation:

$$FCISS = C_{SD} + \sum_{i=1}^{n} N_{SE,i} \cdot C_{SE,i}$$

(2)

Here, the first term C_{SD} is cost for safety design (i.e. cost related to safety classification, safety requirements and design specification, detailed design and engineering, factory acceptance test or pre-start up acceptance test and start up and correction). Table 3.6 gives the typical cost elements included in C_{SD} calculations. The second term $\sum_{i=1}^{n} N_{SE,i} \cdot C_{SE,i}$ s the sum of the purchase cost of safety equipment. Here CSE,i is the purchase cost of equipment "i" and NSE,i is the number (count) of that equipment. The costs for these devices are based on the recent detailed survey of available costs from various suppliers conducted by Khan and Annyotte (2004), however in the module the user has the possibility to enter the present market costs.

Table 6. Typical cost elements included in CSD of safety system cost calculations

Safety classification cost e.g SIL determination cost	C_{SIL}
Safety requirements and design specifications (SRS) cost	C_{SRS}
Detailed design and engineering cost	C_{DE}
!Miscellaneous Cost:	C_{ME}
Initial training cost	C_{TC}
Factory acceptance test (FAT)/Installation/Pre-startup acceptance test (PSAT) cost	C_{FAT}
Startup and correction cost	C_{SCC}
$C_{SD} = C_{SIL} + C_{SRS} + C_{DE} + C_{ME} + C_{TC} + C_{FAT} + C_{SCC}$	

!power, wiring, junction boxes, operators interface cost

ii   Fixed capital investment (FCI2)

Then, extended fixed capital investment is calculated by adding FCISS to FCI1.

$$FCI2 = FCI1 + FCISS \qquad (3)$$

iii.   Total annualized cost (TAC2)

So, the extended total annualized cost will be calculated using extended fixed capital investment.

$$TAC2 = d \cdot (FCI2) + OC1 \qquad (4)$$

Maintenance and repair cost of safety system should also be included in this calculation. But in this economic module these cost elements are not considered.

iv. Risk cost 1 (RC1)

Risk cost (RC1), which is the sum of property risk cost due to asset loss (PRC), health risk cost due to human health loss (HRC) and environmental risk cost due to environmental damage (ERC). The relations for calculation of these costs used in the module are:

Property risk cost due to asset lost (PRC) is the cost incurred due to lost of physical assets such as damage to property, loss of equipment due to accident/scenario and calculated by the equation below:

$$PRC = \sum_{i=1}^{n} \dot{F}_{A,i} \cdot A_{D,i} \cdot C_{A,i} \cdot t_{op} + \sum_{j=1}^{n} \dot{F}_{I,j} \cdot C_{D,j} \cdot t_{op}$$

$$(5)$$

$\dot{F}_{A,i}$ is frequency of occurring the hazardous accident, $A_{D,i}$, is damage area due to that accident, $C_{A,i}$ is the asset cost per unit area, t_{op} is total operation time, $\dot{F}_{I,j}$ is incident occurring frequency and $C_{D,j}$ is incident damage cost.

Health risk cost due to human health lost (HRC) is the cost of fatality and/or injury due to the accident scenario under study.

$$HRC = \sum_{i=1}^{n} \dot{F}_{A,i} \cdot N_{Peop,eff} \cdot C_{H,life} \cdot t_{op}$$

$$(6)$$

Here, $N_{peop,eff}$ is the number of person affected due to accident and is equal to $N_{peop\,eff}$, $= \times$ pop $\tilde{A}$. Where POP is the population around the area of accident and $\wp$ is the population distribution factor ($\wp$ is 1 if population is uniform distributed (maximum value) and $\wp$ is 0.2 if population is localized and away from the area of accident (minimum value)) and $C_{H,life}$ is dollar value of human life or health. Though attempts to put value on human life have caused criticism and it changes from place to place. But a value for this can be obtained by dividing the annual gross national product by the annual number of births or by estimating how much money the person would have earned if not killed by the accident (Tweeddale 2003). A value for cost of loss of lives, marginal cost to avert the fatality, for the highest category of involuntariness risk 14 x 106 $ is used in this work (Passman, H.J. et al. 2003).

Environmental risk cost due to environmental damage is the cost incurred due to environmental damage.

$$ERC = \sum_{i=1}^{n} \dot{F}_{ED} \cdot A_{ED,i} \cdot C_{ED,i} \cdot t_{op}$$

(7)

Where AED i, is the environmental damage area due to scenario "i", FED & is the frequency of release of material to environment and $C_{ED\,i}$, is the environmental damage cost per unit area. so the sum of these three risk costs gives:

$$RC1 = PRC + HRC + ERC$$

(8)

v. Risk cost 2 (RC2) Risk cost 2 (RC2), which is the sum of risk cost1 (RC1) and production loss risk cost (PLRC), accounts for accident damage risk cost. Here, production loss risk cost due to asset damage (PLRC) accounts for the cost due to the production loss because of accident and given by:

$$PLRC = \sum_{i=1}^{n} \dot{F}_{A,i} \cdot t_d \cdot \dot{C}_p \cdot t_{op}$$

(9)

Where PLRC is the production loss risk cost, td is the time lost due to accident and $\dot{C}_p$ is the production loss value in $/h. Thus Risk cost 2 (RC2) is

$$RC2 = RC1 + PLRC$$

vi.   Risk cost 3 (RC3)

Risk cost 3 (RC3), which is sum of process interruption cost due the spurious trip of the safety system and process interruption cost because of safe shut down to avoid from accident, accounts for incident damage risk cost and calculated as follow:

$$RC3 = (\sum_{i=1}^{n} \dot{F}_S^{trip} \cdot t_{trip} + \dot{F}_R^{trip} \cdot t_{dR}) \cdot \dot{C}_p \cdot t_{op}$$

(10)

Here, $\dot{F}_S^{trip}$ is spurious trip frequency, $\dot{F}_R^{trip}$ is safe shut down frequency when trip system demand arises, ttrip is down time due to spurious trip and t_{dR} is down time to safe shut down when trip system demand arises.

vii.   Total Risk cost (TRC)

Total Risk cost (TRC) is the sum of all risk costs:

$$TRC = RC2 + RC3$$

(11)

Total risk cost can be annualized by dividing it with total operation time (t_{op}):

$$TAC_{risk} = TRC / t_{op}$$

viii.   Extended Cost (ECC)

Extended cost calculations (ECC) is Life cycle related cost and calculated as follow :

$$ECC = FCI2 + PVC \qquad (12)$$

Here, PVC is present value of the annual costs (OC1, TAC_{risk}) and calculated as follow:

$$PVC = (OC1 + TAC_{risk} - Insurance\ cos t) \cdot \frac{1 - (1 + R)^{-t_{ly}}}{R}$$

R is the present interest rate and t_{ly} is the number of years (predicted life of system).

Besides, the cost elements mentioned above in Extended cost calculation section, the other elements such as warranty/insurance cost, lost of image and prestige cost, market lost cost should also be considered but quantification of these elements is still almost impossible.

Environment Module

Environment module consists of four steps and introduced an environmental performance index (EPI1) for evaluation of environmental performance and environmental pollution index (EPI2) as environmental objective to be integrated along with economics. The environmental performance index (EPI1) is calculated by combining total PEI based on WAR algorithm[16,17], resource depletion, energy conservation and fugitive emission rates while environmental pollution index (EPI2) is calculated by combining total PEI based on WAR algorithm and fugitive emissions because in this case other factors like resource depletion and energy consumption will be integrated in economic module or objective function. The Analytic hierarchy process (AHP) is used as multicriteria decision analysis tool for combining these different impacts and determination of weighting factors of individual impact categories in total PEI and later on in environmental performance index (EPI1) and environmental pollution index (EPI2) calculations. The module is developed using Microsoft Visual Basic 6.0 and WAR GUI (WAR graphical user interface) is integrated in the user plate form. The steps are:

Step I : Problem definition and data gathering

Step II : Individual impact categories calculation

Step III: Determination of weighting factors

Step IV: Environmental performance index calculation

Figure 10 shows the simplified block diagram of environment module and tasks to be performed.

Step I: Problem Definition and Data Gathering

The primary task in step 1 is problem framing and scope definition. Information such as material and energy balance information, process conditions, process technology and nature of used materials/chemicals should be retrieved from process module. Process flow diagram is to be re-examined for identification of additional waste and emission streams. Collect additional data and information for environment evaluation to fill gaps. As sources of emissions such as fugitive emission sources, venting of equipment, periodic equipment cleaning, incomplete separations etc. are often missing in process so process is analyzed to identify these sources.

Step II: Individual Impact Categories Calculation (Potential Environmental Impact Calculations Based On War Algorithm)

The software WAR GUI (waste reduction algorithm graphical user interface) from the US Environmental Protection Agency is used to calculate individual potential environmental impacts. The generalized formula based on WAR algorithm for calculating individual PEI is given in equation 13.

$$PEI_L = (\dot{M}_b \cdot \sum_{k}^{Comps} x_{kb} \cdot \psi_{kL} + \dot{Q}_r \cdot \psi_L^E) / \dot{M}_p \qquad [Impact / kg\ product] \tag{13}$$

Where PEIL is the potential environmental impact of category L, $\dot{M}_b$ is mass flow rate of base (effluent) stream, x_{kb} is the mass fraction of component k in the base stream, ψ_{kL} is the normalized impact score of chemical k for category L, $\dot{Q}_r$ is energy rate supplied for separation and ψ_L^E is the normalized impact score of category L due to energy. The sensitivity analysis results of individual potential environmental impact with respect to optimization variables should also be performed.

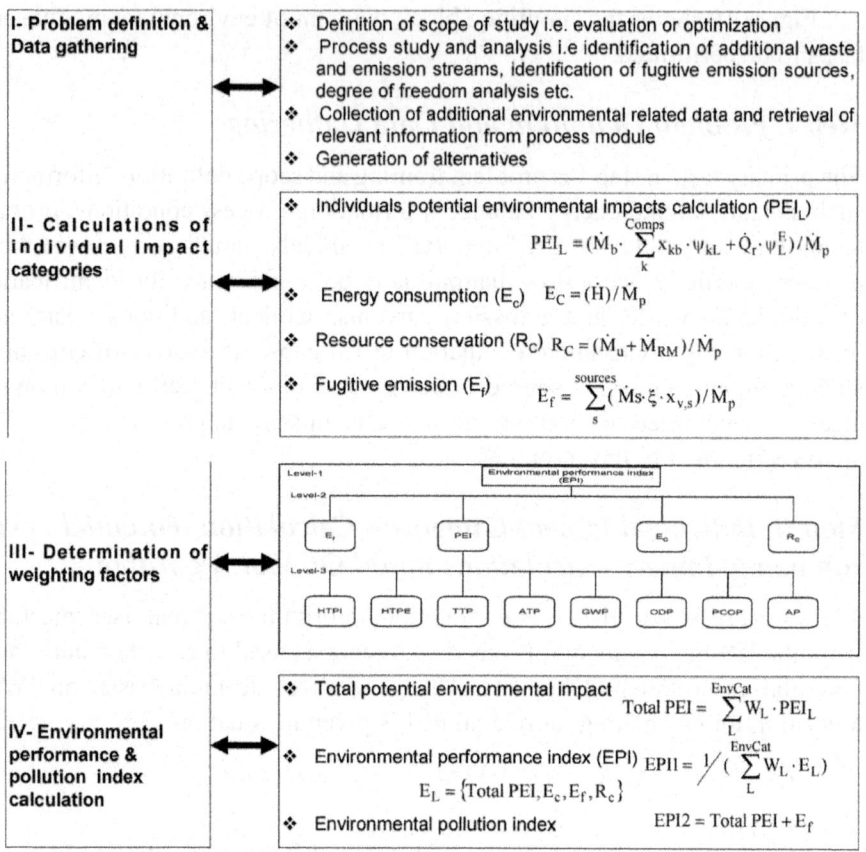

Figure. 10. Simplified block diagram of environment module.

Energy Consumption Factor (E_c)

Energy consumption factor refers the total amount of energy consumed in the process per unit of product and is calculated as follow:

$$E_C = (\dot{H}) / \dot{M}_p \qquad [kJ \, / \, kg \; product] \qquad (14)$$

Here $\dot{H} = \dot{M}_{steam} \cdot \hat{h}_{steam} + E_E$ where $\dot{M}_{steam}$ is the mass flow rate of steam [kg/h], $\hat{h}_{steam}$ is the enthalpy of steam per kg [KJ/kg], EE is electrical energy consumed per unit time [KJ / h] and $\dot{M}_p$ is product rate [kg/h].

The sensitivity analysis of this factor with respect to optimization variables should also be performed.

Resource Conservation Factor (RC)

The resource consumption refers all needed raw materials and utilities used and given by:

$$R_C = (\dot{M}_u + \dot{M}_{RM}) / \dot{M}_p \qquad [\text{kg / kg product}] \qquad (15)$$

Where R_C is the resource conservation factor, $\dot{M}_u$ is utilities consumption rate, $\dot{M}_{RM}$ is raw material consumption rate.

Fugitive Emission Factor (E_f)

Fugitive emissions are unplanned or unmanaged, continuous or intermittent releases from unsealed sources such as storage tank vents, valves, pump seals, flanges, compressors, sampling connections, open ended lines etc and any other non point air emissions. These sources are large in number and difficult to identify. These emission rates depends on factors such as the age and quality of components, specific inspection and maintenance procedures, equipment design and standards of installation, specific process temperatures and pressures, number and type of sources and operational management commitment[18]. However, four basic approaches for estimating emissions from equipment leaks in a specific processing unit, in order of increasing refinement, in use are:

- Average emission factor approach
- Screening ranges approach
- EPA correlation approach
- Unit-specific correlation approach

All these approaches require some data collection, data analysis and/or statistical evaluation. On the other hand, using fundamental design / engineering calculations for accurate fugitive emission estimations for each source present in the process industry are difficult due to:

- large number and type of fugitive emission sources
- dependence of emission rates on other factors along with design and operating conditions e.g. installation standards, inspection and maintenance procedure etc.

As focus in this work is to integrate fuggitve emissions into environmental performance evaluation and optimization objectives so average emission factor approach giving a bit over estimates are used. Average emission factors for estimating fugitive emissions from fugitive sources found in synthetic organic chemical manufacturing industries operations (SOCMI) obtained from the US

Environmental Protection Agency L & E Databases are used. The relation used in this work for calculation of fugitive emissions is:

$$E_f = \sum_{s}^{sources} (\dot{M}s \cdot \xi \cdot x_{v,s}) / \dot{M}_p \ \text{[kg/ kg product]}$$

(16)

Here E_f is fugitive emission factor per unit of product, $\dot{M}_s$ mass flow rate through the source 's', x is average emission factor and x_{vs} is mass fraction of volatile component through source 's' and $\dot{M}_p$ is product rate. It is assumed x_{vs} for the process fluids through fugitive sources such as pump seals, valves, flanges and connection is equal to 1, i.e. fluids are composed entirely of volatile compounds.

Step Iii: Determination of Weighting Factors (Application of Multicriteria Decision Analysis Technique)

The integration of these individual impact categories into one index is a hierarchical multicriteria decision analysis problem. The analytic hierarchy process (AHP) is used for this purpose[19] and a computer programme for it is developed in VB 6.0. In this stage, first a hierarchical structure of the problem, which is structured hierarchically similar to a flow chart, is constructed. The overall objective is placed at the top while the criteria and subcriteria are placed below. For example, as shown in figure 11, the overall objective Environmental performance index (EPI1) is placed at the top (level 1), then below (level 2) are criterias Total PEI, Ef, Ec and RC and after this (level 3) sub-criterias as HTPI, HTPE, TTP, ATP, GWP, ODP, PCOP and AP. After this using the numerical scale given in table 2.6, two pairwise comparison matrices (see Table 7 and 8) are constructed for determination of weights for aggregation of individual impact categories of WAR to total PEI and for determination of weights of total PEI, Ef, Ec and RC to Environmental performance index (EPI1).

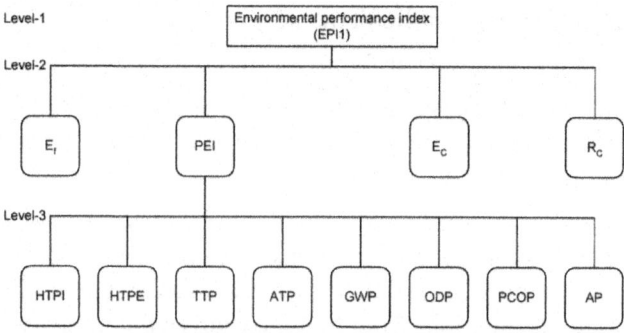

Figure. 11. Hierarchical structuring of multicriteria decision analysis problem for integrating individual environmental impacts.

The right hand upper diagonal information in both matrices is to be provided by the decision maker giving the relative importance of the two criteria using the numerical scale of table 2.6 while the left hand lower diagonal is the reciprocal of the right hand upper diagonal. Once these pair wise comparison matrices are constructed, then developed computer programme using the AHP method, determines the weighting of individual impact categories. The level of inconsistency of decision makers input is checked by consistency ratio before giving the output. Consistency ratio less than 0.1 is good and for ratios greater than 0.1, the input to pair wise matrix should be re-evaluated.

Table 7. Pairwise comparison matrix for individual impact categories at level 3

Pairwise comparison matrix								
	HTPI	**HTPE**	**TTP**	**ATP**	**GWP**	**ODP**	**PCOP**	**AP**
HTPI	1	A12	A13	A14	A15	A16	A17	A18
HTPE		1	A23	A24	A25	A26	A27	A28
TTP			1	A34	A35	A36	A37	A38
ATP				1	A45	A46	A47	A48
GWP					1	A56	A57	A58
ODP						1	A67	A68
PCOP							1	A78
AP								1
W$_L$	W1	W2	W3	W4	W5	W6	W7	W8

Table 8. Pairwise comparison matrix for individual impact categories at level 2

Pairwise comparison matrix			
	PEI	**R$_C$**	**E$_C$**	**E$_f$**
PEI	1	A12	A13	A14
R$_C$		1	A23	A24
E$_C$			1	A34
E$_f$				1
W$_L$	W1	W2	W3	W4

Step IV: Environmental Performance & Pollution Index Calculation

In the final step, first Total PEI is determined by multiplying each impact category values with its relevant weighting factor WL as given below:

$$Total\ PEI = \sum_{L}^{EnvCat} W_L \cdot PEI_L$$

(17)

After calculating Total PEI, Environmental performance index (EPI1) is determined for each alternative by multiplying the values of Total PEI, Ef, EC and RC with its relevant weighting factor WL (table 8) as given below:

$$EPI1 = 1 \Big/ \ (\sum_{L}^{EnvCat} W_L \cdot E_L)$$

(18)

Where $E_L = \{Total\ PEI, E_c, E_f, R_c\}$

and environmental pollution index (EPI2) is calculated as follow

$$EPI2 = Total\ PEI + E_f$$

(19)

The higher value of environmental performance index (EPI1) shows that the process is environmentally better and vice versa. While the higher value of environmental pollution index (EPI2) shows that the environmental performance of process is worse.

Data Manager

The relevant information generated from process module, safety module and environment module for each alternative is transferred to data manager. This information is used to formulate process diagnostic tables and multiobjective decision-making problem formulation. These tables consist of mass input/output table, energy input/output table, capital and utility annual expense summary, environmental impact summary and frequency of occurance of an event and their consequence categories and safety cost.

Stage III: Multiobjective Decision Analysis/ Optimization

The purpose of this layer/stage is to set up multiobjective decision making/ optimization among these conflicting objectives. The aim is to find out the trade-off surface for each alternative and /or complete ranking of alternatives. The calculation loop used for it is shown in figure 12.

In each independent performance module i.e. economic, environment and safety module, relevant information is generated and transferred automatically or manually to data manager for each alternative generated or under study. Before transferring the values of performance objective functions, each objective function is optimized within their independent module such as:

- Process/Economic optimization of each alternative is carried out using SQP optimization algorithm build within Aspen PlusTM.

- The lower and upper limits for Environmental objective functions are calculated using environmental module from the material and energy balance information from process model.

- Safety/risk aspects are optimized in the safety module and information such as hazard occurance frequency, safety cost data (fixed safety system cost, accident and/or incident damage risk cost) for each alternative is transferred to the data manger.

Depending on the case under study or objectives of the study, graphical tool box of MatLab and/or multiobjective optimization technique (goal programming) or multiattribute decision analysis technique (PROMETHEE and/or AHP) is used for multiobjective decision analysis. The Data Manager is linked with MatLab 7.0 via Excel link Toolbox in the integrated interface. Aspen plusTM is linked in the integrated interface via Visual basic 6.0 and Microsoft Excel. AHP technique is also programmed for the cases under study in Visual basic 6.0. The computer realization of these links in the integrated interface is explained in chapter four.

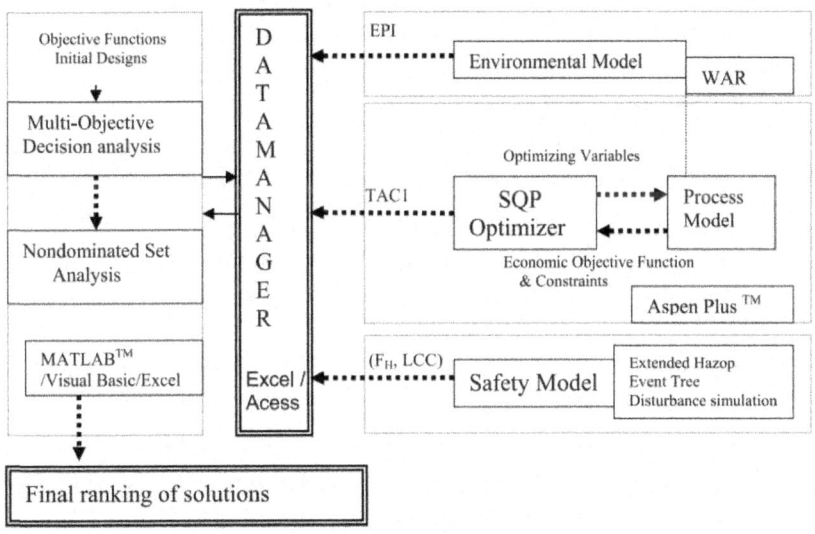

Figure. 12. Calculation loop for multiobjective optimization.

Stage IV: Design Evaluation

The purpose of this layer/stage is to select the best alternative and/or find the complete ranking of alternatives under study based on the results of third stage/layer of the developed methodology. Pareto approach (non dominated analysis) or PROMETHEE is used for this purpose.

REFERENCES

1. Ramzan, N., Naveed, S., Feroze, N. and Witt, W.; "Multicriteria decision analysis for safety and economic achievement using PROMETHEE: A case study" Process Safety Progress (A Journal of Americal Institute of Chemical Engineering), 28(1), 68-83 (2009).

2. Ramzan, N., Degenkolbe, S. and Witt, W.; "Evaluating and improving environmental performance of HC's recovery system: A case study of distillation unit", Chemical Engineering Journal (Journal published by Elsevier), 40(1-3), 201-213(2008). ISSN: 1385-8947.

3. Ramzan, N., Compart, F. and Witt, W.; "Methodology for the generation and evaluation of safety system alternatives based on extended Hazop" Process Safety Progress (A Journal of Americal Institute of Chemical Engineering), 26(1),35-42 (2007).

4. Ramzan, N., Compart, F. and Witt, W.; "Application of extended Hazop and event tree analysis for investigating operational failures and safety optimization of distillation column unit" Process Safety Progress (A Journal of Americal Institute of Chemical Engineering), 26(3),248-257 (2007).

5. Ramzan, N., Witt, W.; "Multiobjective optimization in distillation unit: A case study", The Canadian Journal of Chemical Engineering, 84(5), 604-613(2006).

6. Seader et al. (1999), Process design principles-Synthesis, Analysis, and Evaluation, John Wiley & Sons Inc. New York, 338-370.

7. Nawaz, Z. Mahmood, Z. (2006) Importance Modeling, Simulation and Optimization in Chemical process Design, The Pakistan Engineer (Journal of Institute of Engineers Pakistan), 38-39.

8. Cameron, I., Raman, R. (2005), Process Systems Risk Management, Elsevier Academic Press, NY, ISBN 0-12-156932-2

9. Crowl, D.A., Louvar, J.F. (1999), Chemical Process Safety: Fundamentals with applications," Prentice Hall, New York.

10. Kletz, T.A. (1997), Hazop-past and future, Reliability engineering and system safety, 55, 263-266.

11. Lees, F.P. (1996), Loss prevention in CPI, Butterworth's, London, UK.

12. Peters, M. S. and K.D. Timmerhaus (1991), Plant Design and Economics for Chemical Engineers, Ed.2nd, McGraw-Hill, New York, 90-145.

13. Douglas, J. (1998), Conceptual design of chemical processes, McGraw Hill Inc.

14. Guthrie, K.M. (1969), Data and techniques for preliminary capital cost estimating, Chem. Eng., 114-1421.

15. Guthrie, K.M. (1974), Process plant estimating, evaluation and control, Craftsman, Solano Beach, CA.

16. Cabezas, H. Bare, C. & Mallick, K. (1999), Pollution prevention with chemical process simulators: the generalized waste reduction (WAR) algorithm-full version. Computers and Chemical Engineering, 23,623-634.

17. Cabezas, H. Bare, C. & Mallick, K. (1997), Pollution prevention with chemical process simulators: the generalized waste reduction (WAR) algorithm, Computers and Chemical Engineering, 21s, s305-s310.

18. Dimian, A. C. (2003), Integrated design and simulation of chemical processes, 1st Eds., Elsevier Netherlands, 1-30,113-134.

19. Dev, P.K. (2004), Analytic hierarchy process helps evaluate project in Indian oil pipelines industry, International journal of operation and production management, 24(6), 588-604

Chapter 11

PREDICTING CHEMICAL TOXICITY EFFECTS BASED ON CHEMICAL-CHEMICAL INTERACTIONS

Lei Chen[2], Jing Lu[3], Jian Zhang[4], Kai-Rui Feng[5], Ming-Yue Zheng[3], Yu-Dong Cai[1]

[1] Institute of Systems Biology, Shanghai University, Shanghai, China

[2] College of Information Engineering, Shanghai Maritime University, Shanghai, China

[3] Drug Discovery and Design Center (DDDC), Shanghai Institute of Materia Medica, Shanghai, China

[4] Department of Ophthalmology, Shanghai First People's Hospital Affiliated to Shanghai Jiaotong University, Shanghai, China

[5] Simcyp Limited, Blades Enterprise Centre, Sheffield, United Kingdom

ABSTRACT

Toxicity is a major contributor to high attrition rates of new chemical entities in drug discoveries. In this study, an order-classifier was built to predict a series of toxic effects based on data concerning chemical-chemical interactions under the assumption that interactive compounds are more likely to share similar toxicity profiles. According to their interaction confidence scores, the order from the most likely toxicity to the least was obtained for each compound. Ten test groups, each of them containing one training dataset and one test dataset, were constructed from a benchmark dataset consisting of 17,233 compounds. By a Jackknife test on each of these test groups, the 1st order prediction accuracies of the training dataset and the test dataset were all approximately 79.50%, substantially higher than the rate of 25.43% achieved by random guesses. Encouraged by the promising results, we expect that our method will become a useful tool in screening out drugs with high toxicity.

INTRODUCTION

Toxicity is a key cause of late-stage failures in drug discovery. Even some approved drugs such as Phenacetin [1] and Troglitazone [2] have been withdrawn from the market because of unexpected toxicities that were not detected during Phase III clinical trials. Thus, early toxicology data on compounds are needed to reduce R&D costs. Evaluating toxicity and assessing risks of diverse chemicals require comprehensive experimental testing against a broad spectrum of toxicity end points. These tests can cost millions of dollars, involving several thousand animals, and take many years to complete. As a result, very few chemicals have undergone the degree of testing needed to support accurate health risk assessments or meet regulatory requirements for drug approval. In recent years, the number of synthetic compounds has surged with the advance of combinatorial chemistry, and accordingly large quantities of toxicity data are urgently demanded.

Recently, particular interest has been raised to apply fast and cost-effective *in silico* toxicological models to supplement those *in vitro* and *in vivo* testing. These models require high quality toxicity data for a large set of structurally diverse drug candidates. Accelrys Toxicity is a database of toxicity information compiled from the open scientific literature [3] and containing toxicological data for approximately 0.17 million chemicals. This database is of great value for investigating the pharmacokinetic properties, metabolism and potential toxicities of compounds. Six types of toxicity data are collected in the database:

1) Acute Toxicity;

2) Mutagenicity;

3) Tumorigenicity;

4) Skin and Eye Irritation;

5) Reproductive Effects; and

6) Multiple Dose Effects.

It should be noted that these categories have multiple and overlapping mechanisms of toxic action and each category represents only specific types of experiments. The combination of these experimental results may help define the overall safety profile of a compound. However, this kind of databases only provides toxicological information for recorded compounds, not for new ones. It would be valuable to accurately predict toxicities of a new compound based on the information available for recorded compounds. In order to meet the demand, there is a drive to develop quick, reliable, and non-animal-involved prediction methods, *e.g.* using structure-activity relationships (SARs) to predict drugs toxicities.

Currently, most toxicological SAR models belong to binary classifiers, which only predict compounds to be toxic or non-toxic within a single toxicity class [4], [5]. It is desired to modify the strategy to predict a series of toxicity effects. In this study, we chose to build a multiclass model [6], [7] to predict six categories of toxicity using the Accelrys Toxicity database instead of only one or two toxicity endpoints. However, the quadratic optimization problem in multiclass models is difficult to solve. Thus, many previous multiclass approaches tended to decompose a multiclass problem into multiple independent binary classifications. Investigators built a set of binary classifiers, such as the model of Dietterich et al [7], each classifier distinguishing only one of the classes from the others. Although this greatly simplifies the problem, such an approach cannot provide order prediction information for the query compounds. That is, it can only predict whether the query compound has some toxicity end points, but cannot determine which is the most likely toxicity, or even the order of toxicity end points by toxicity likelihoods.

In recent years, the assessment of protein-protein interactions has been widely used to predict many attributes of proteins [8], [9], [10], [11]. Furthermore, multiclass predictions of protein attributes have become more common [12], [13], [14]. These methods and their results show that interactive proteins tend to share the same functions with higher probability than do non-interactive ones. Likewise, it is reasonable to expect that interactive compounds are also more likely to share common functions as indicated by some pioneer studies [15], [16]. Thus, toxicity, as part of the biological functions of compounds, should follow the same rule. Moreover, based on a previous work on the Anatomical Therapeutic Chemical (ATC) classification of drugs [16], compared to the SAR models based on physicochemical descriptors or structural alerts, a model based on chemical-chemical interactions can rank the order of the predictions more easily and yield better prediction results. In our study, we attempt to quantify chemical-chemical interactions for each pair of interactive compounds, and obtain the confidence scores of the interactions by which the toxicity end points were ordered. Briefly, compounds of seven categories including six categories of toxicity plus non-toxicity were collected. The interactive compounds of each query compound were identified utilizing STITCH (Search tool for interactions of chemicals) [17], [18]. Then, the score of each class of the query compound was obtained from the confidence scores of interactions between the query compound and its interactive compounds using the toxicity profile of the interactive compounds. Finally, the prediction quality of the model was evaluated using the Jackknife test through ten test groups. Each of these was constructed from the benchmark dataset and contained one training dataset and one external test dataset. Details are described in the following sections.

MATERIALS AND METHODS

Benchmark Dataset

We obtained a total of 171,266 compounds from the Accelrys Toxicity Database 2011.4 [19], which had at least one toxicity effect belonging to the following six categories:

1) Acute Toxicity;

2) Mutagenicity;

3) Tumorigenicity;

4) Skin and Eye Irritation;

5) Reproductive Effects;

6) Multiple Dose Effects.

Based on compound toxicity, these compounds are allocated to the 6 categories, allowing multiple assignments. In addition, 2,871 "non-toxic" compounds including FDA-approved drugs from DrugBank [20] and endogenic metabolites from the Human Metabolome database (HMDB) [21] were collected and labeled as a negative class. For convenience, the 'non-toxic set' is regarded as the 7th category of compound toxicity. Due to lack of chemical-chemical interaction information in STITCH [17], [18], some compounds cannot be investigated by this approach. After excluding these compounds, a benchmark dataset S consisting of 17,233 compounds was retrieved, of which 16,587 were toxic and 646 were non-toxic. These compounds are classified into 7 categories of compound toxicity. Shown in **Table 1** is the distribution of compounds in each category.

Table 1. Distribution of compounds in each category of compound toxicity

Tag	Toxicity	Total
T_1	Acute Toxicity	12,633
T_2	Mutagenicity	6,110
T_3	Tumorigenicity	2,293
T_4	Skin and Eye Irritation	2,353
T_5	Reproductive Effects	2,501
T_6	Multiple Dose Effects	4,198
T_7	Non-toxicity	646
Total	–	30,734

doi:10.1371/journal.pone.0056517.t001

It is observed from **Table 1** that the sum of the number of compounds in all the 7 categories is much larger than the number of compounds, indicating that some compounds are allocated to more than one category of toxicity. Of the 17,233 compounds in the benchmark dataset, 10,151 compounds belong to only one category of toxicity, 3,475 compounds belong to two categories of toxicity, while others belong to 3–5 categories of toxicity and no compounds belong to more than five categories of toxicity - refer to **Figure 1** for a plot of the number of compounds against the number of categories of toxicity. Thus, prediction of compound toxicity is a multi-label classification problem. Like the case of processing proteins or compounds with multiple attributes [15], [16], [22], the proposed method would provide a series of candidate toxicities, ranging from the most to the least likely, instead of presenting only the most likely one.

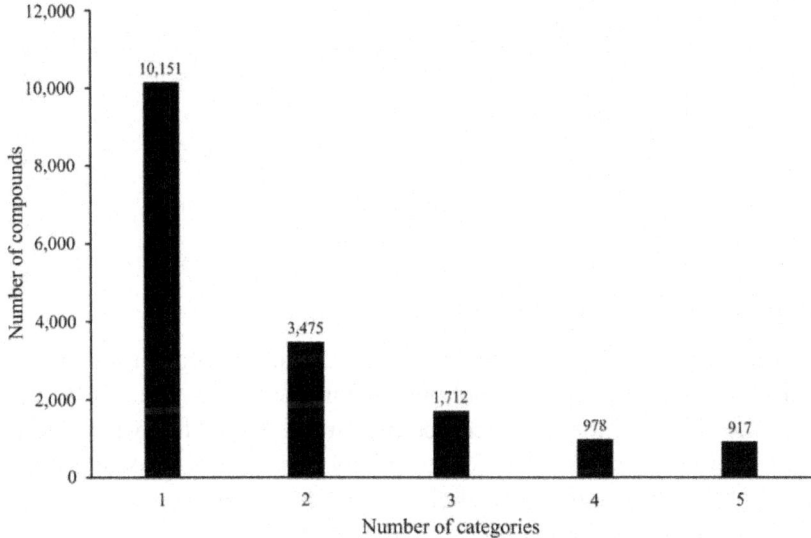

Figure 1. The number of compounds plotted against the number of categories in the benchmark dataset.

To sufficiently evaluate the prediction method described in the following section, we constructed 10 test groups, denoted by $TG_1, TG_2, \ldots, TG_{10}$, respectively. In each test group $TG_i (1 \leq i \leq 10)$, there is one training dataset $S_{tr}^{(i)}$ and one test dataset $S_{te}^{(i)}$, i.e., $TG_i = \langle S_{tr}^{(i)}, S_{te}^{(i)} \rangle$, where the test dataset consisted of 1,723 compounds which were randomly selected from S, while the training dataset contained the remaining 15,510 samples in S, i.e., $S = S_{tr}^{(i)} \cup S_{te}^{(i)}$ for each $1 \leq i \leq 10$. It is necessary to point out that, in each test group, the portion of the data in each class of the test dataset is roughly the same as that of the training dataset. Shown in **Table 2** is the distribution of compounds in training and test datasets of each test group.

Table 2. Distribution of compounds in training and test datasets of each test group

Tag	TG_1		TG_2		TG_3		TG_4		TG_5	
T_1	11,382	1,251	11,387	1,246	11,351	1,282	11,364	1,269	11,385	1,248
T_2	5,475	635	5,476	634	5,529	581	5,492	618	5,491	619
T_3	2,065	228	2,065	228	2,063	230	2,063	230	2,056	237
T_4	2,102	251	2,102	251	2,115	238	2,112	241	2,093	260
T_5	2,235	266	2,235	266	2,260	241	2,255	246	2,235	266
T_6	3,747	451	3,749	449	3,777	421	3,784	414	3,799	399
T_7	582	64	577	69	586	60	582	64	583	63
Total	27,588	3,146	27,591	3,143	27,681	3,053	27,652	3,082	27,642	3,092

Tag	TG_6		TG_7		TG_8		TG_9		TG_{10}	
T_1	11,367	1,266	11,395	1,238	11,369	1,264	11,374	1,259	11,353	1,280
T_2	5,489	621	5,500	610	5,492	618	5,497	613	5,506	604
T_3	2,075	218	2,067	226	2,070	223	2,043	250	2,070	223
T_4	2,123	230	2,125	228	2,135	218	2,102	251	2,133	220
T_5	2,244	257	2,243	258	2,236	265	2,258	243	2,234	267
T_6	3,762	436	3,750	448	3,772	426	3,777	421	3,755	443
T_7	583	63	587	59	579	67	569	77	584	62
Total	27,643	3,091	27,667	3,067	27,653	3,081	27,620	3,114	27,635	3,099

doi:10.1371/journal.pone.0056517.t002

Chemical-Chemical Interactions

It is known that two proteins that can interact with each other are more likely to share common biological functions than non-interactive ones [8], [9], [10], [11]. Likewise, two interactive compounds are also more likely to share similar biological functions [15], [16]. Since toxicity is one of a compound's properties and functions, utilizing chemical-chemical interactions to identify compound toxicity is deemed to be feasible.

The data for chemical-chemical interactions were retrieved from STITCH (chemical_chemical.links.detailed.v3.0.tsv.gz, http://stitch.embl.de/cgi/show_download_page.)[17], a well-known database including known and predicted interactions of chemicals and proteins collected from experiments, literature or other reliable sources. In the obtained file, the interaction unit contains two compounds and five kinds of scores with titles "Similarity", "Experimental", "Database", "Textmining" and "Combined_score". The last kind of score was used here to indicate the interactivity of two compounds, *i.e.*, two compounds with "Combined_score" greater than zero were deemed interactive compounds, because the last kind of score integrates the information of the other kinds of scores. Thus, the considered interactive compounds in this study contain the following three categories:

1) those participating in the same reactions;

2) those sharing similar structures or activities and

3) those with literature associations [17].

It is known that these categories correspond to the following three facts: (I) compounds involved in the same reactions occupy the same biological pathways; (II) compounds with similar structures or activities are likely to share similar functions, thereby occupying the same pathways with high probability; (III) the co-occurrence of two compounds, as noted in many studies, indicates some direct or indirect relationships, suggesting that they have the potential to share the same pathways. On the other hand, compounds in the same biological pathways always induce similar side effects, thereby having similar toxicity effects. Accordingly, it is reasonable to suppose that interactive compounds tend to have similar toxicity effects.

The value of the "Combined_score" of two interactive compounds indicates the likelihood that they can interact, *i.e.*, two interactive compounds with high "Combined_score" can interact with high probability. Thus, this score is also termed a confidence score in this study. For two compounds c_1 and c_2, let us denote the confidence score of an interaction between them by $Q(c_1, c_2)$. Specifically, if there is no interaction information between c_1 and c_2 based on the current records in STITCH, their interaction confidence score is assigned zero, *i.e.*, $Q(c_1, c_2) = 0$. In this study, 323,432 interaction units, *i.e.*, 323,432 pairs of compounds with confidence scores greater than 0, were used to predict compound toxicity.

Prediction Method

As is mentioned in the above section, interactive compounds are more likely to have common toxicity. Accordingly, the toxicities of a query compound can be identified according to its interactive compounds.

For convenience, let T_1, T_2, ..., T_7 denote the seven categories of toxicity, where T_1 denotes "Acute Toxicity", T_2 "Mutagenicity", and so forth (see column 1 and 2 of **Table 1**). Suppose that there are n compounds in the training dataset, that is c_1, c_2, ..., c_n, the toxicity of a compound c_i in the training dataset is formulated as

$$T(c_i) = [t_{i,1}, t_{i,2}, \ldots, t_{i,7}](i = 1, 2, \ldots, n) \tag{1}$$

where

$$t_{i,j} = \begin{cases} 1 & \text{If } c_i \text{ has toxicity } T_j \\ 0 & \text{Otherwise} \end{cases} \tag{2}$$

Given a query compound c_q, its toxicity is predicted not only by its interactive compounds but also by the confidence scores of their interactions. The score indicating that the query compound c_q has toxicity T_j is calculated by

$$\Theta(c_q \mapsto T_j) = \sum_{i=1}^{n} Q(c_i, c_q) \cdot t_{i,j}, j = 1,2,3,4,5,6,7 \tag{3}$$

The high score $\Theta(c_q \mapsto T_j)$ means that there are many interactive compounds of c_q in the training dataset that have toxicity T_j or some interactions between c_q and its interactive compounds having toxicity T_j are labeled by high confidence scores. In view of this, the greater the score $\Theta(c_q \mapsto T_j)$, the more likely that the compound c_q has toxicity T_j. In particular, if $\Theta(c_q \mapsto T_j)$ for some j, it is indicated that the probability that the query c_q having the j-th category of toxicity is zero because there are no interactive compounds of c_q in the training dataset that have toxicity T_j.

Since this is a multi-label classification problem, *i.e.*, some compounds have more than one category of toxicity. A prediction method only providing the most likely toxicity is not an optimal choice. Thus, our method is valuable in that it can provide a series of candidate toxicities for a query compound, ranging from the most likely to the least likely. For example, if the results obtained from **Eq. 3** are

$$\Theta(c_q \mapsto T_3) \geq \Theta(c_q \mapsto T_1) \geq \Theta(c_q \mapsto T_6) > 0 \tag{4}$$

it can be interpreted to mean that there are three candidate toxicities for the query compound c_q, and the most likely toxicity for c_q is T_3 ("Tumorigenicity", cf. **Table 1**), followed by T_1 ("Acute Toxicity") and T_6 ("Multiple Dose Effects"). In addition, T_3 is called the 1st order prediction, T_1 the 2nd order prediction, and so forth.

Jackknife Test

The Jackknife test [16] is often used to examine the performance of various predictors, because it can always provide a unique prediction result for a given dataset. It has been widely used by investigators to evaluate their predictors [23], [24], [25], [26], [27], [28], [29], [30], [31],[32], [33]. During the test, each sample in the training dataset is singled out one-by-one and tested by the predictor trained by the other samples. Thus, each sample is tested exactly once.

Accuracy Measurement

The j-th order prediction accuracy is calculated by the following formula [15], [16]:

$$\Gamma_j = \frac{CT_j}{N} j = 1,2,3,4,5,6,7 \tag{5}$$

where CT_j denotes the number of compounds whose j-th order prediction is one of its true toxicities, and N denotes the total number of compounds in the dataset. If a prediction method can obtain high Γ_j with small j and low Γ_j with large j, it implies that the method arranges the candidate toxicities well. Among them, the 1st order prediction accuracy is the most important indicator of good or bad performance.

Although the seven prediction accuracies can be obtained by **Eq. 5**, none of them provides the overall prediction accuracy. In view of this, we employ another measurement that calculates the proportion of true toxicities of the first m predictions. It can be calculated as follows [16]:

$$\Delta_m = \frac{\sum_{i=1}^{N} S_{i,m}}{\sum_{i=1}^{N} N_i} \tag{6}$$

where $S_{i,m}$ represents the number of the correct predictions of the i-th compound among its first m predictions, and N_i represents the number of toxicities that the i-th compound has. Since different compounds may have different numbers of toxicities, the parameter m in **Eq. 6** is usually taken as the smallest integer no less than the average number of toxicities in the dataset,

which can be computed by

$$M = \frac{\sum_{i=1}^{N} N_i}{N} \tag{7}$$

where $m = \lceil M \rceil$. Obviously, a larger Δ_m implies better prediction performance by the method for the identification of compound toxicity.

RESULTS

As described in the Section "Benchmark dataset", 10 test groups were constructed to evaluate the method described in Section "Prediction method". In each test group, there were one training dataset consisting of 15,510 compounds and one test dataset containing 1,723 compounds. The predicted results for each test group obtained by the proposed method are as follows.

Performance of the Method on the Training Dataset

For the 15,510 compounds in each training dataset $S_{tr}^{(i)}(1 \leq i \leq 10)$, we conducted the prediction and evaluated its performance by the Jackknife test. Listed in the column with title $S_{tr}^{(i)}$ of **Table 3** are seven prediction accuracies, calculated by **Eq. 5**, for training dataset $S_{tr}^{(i)}$, from which we can see that the 1st order prediction accuracies were all around 79.50%, where the maximum was 79.57%, while the minimum was 79.23%; the 2nd order ones were all around 37.30%. It is indicated that the proposed method is very stable. It is also observed from the corresponding columns of **Table 3** that the accuracies followed a descending trend when increasing the order number, indicating that the method sorted the candidate toxicities quite well for the compounds in each training dataset $S_{tr}^{(i)}(1 \leq i \leq 10)$. The average numbers of toxicities for compounds in each training dataset $S_{tr}^{(i)}$ were about 1.78 according to **Eq. 7**, i.e., $M=1.78$. It is noteworthy that if one predicts compound toxicity by random guesses, the average success rate would be only 25.43% (1.78/7), which is much lower than each of the 1st order prediction accuracies by our method. To evaluate the prediction accuracy by the method more thoroughly, **Eq. 6** was calculated by taking $m=2$, i.e., we considered the first two predictions for each compound in $S_{tr}^{(i)}(1 \leq i \leq 10)$ to see the proportions of true toxicities covered by these predictions. These proportions are shown in column 2 of **Table 4**, from which we can see that they were all about 65.50%, where the maximum was 65.61% while the minimum was 65.32%. Thus, it is indicated once again that our method is reliable.

Table 3. Prediction accuracies obtained by the method as applied to training and test datasets of each test group

Prediction Order	TG_1		TG_2		TG_3		TG_4		TG_5	
1	79.40%	79.69%	79.45%	79.28%	79.23%	80.62%	79.28%	79.45%	79.30%	79.34%
2	37.16%	38.42%	37.14%	38.24%	37.54%	37.20%	37.17%	38.31%	37.40%	36.16%
3	22.18%	23.16%	22.20%	22.87%	22.32%	21.65%	22.29%	22.63%	22.53%	22.87%
4	15.45%	16.66%	15.49%	16.77%	16.35%	14.86%	15.46%	16.13%	15.41%	15.55%
5	11.06%	11.61%	11.04%	11.49%	11.00%	10.85%	10.88%	10.16%	10.95%	11.20%
6	6.92%	7.25%	6.84%	7.89%	7.23%	5.86%	6.99%	6.56%	6.85%	7.84%
7	1.21%	1.33%	1.22%	1.04%	1.27%	1.51%	1.39%	1.45%	1.26%	1.68%

Prediction Order	TG_6		TG_7		TG_8		TG_9		TG_{10}	
	$S_{tr}^{(6)}$	$S_{te}^{(6)}$	$S_{tr}^{(7)}$	$S_{te}^{(7)}$	$S_{tr}^{(8)}$	$S_{te}^{(8)}$	$S_{tr}^{(9)}$	$S_{te}^{(9)}$	$S_{tr}^{(10)}$	$S_{te}^{(10)}$
1	79.57%	80.15%	79.36%	79.98%	79.45%	79.05%	79.52%	79.80%	79.46%	79.34%
2	37.11%	37.72%	37.57%	36.10%	37.21%	38.65%	37.32%	35.98%	37.44%	37.20%
3	22.57%	22.29%	22.30%	23.39%	22.23%	24.03%	22.46%	23.33%	22.42%	22.93%
4	15.31%	15.90%	15.36%	15.55%	15.52%	14.74%	15.40%	16.25%	15.36%	16.37%
5	10.93%	10.45%	10.95%	11.55%	11.08%	10.10%	10.74%	11.55%	10.87%	10.74%
6	7.00%	6.56%	7.00%	6.62%	7.16%	5.86%	6.76%	7.78%	6.97%	7.25%
7	1.25%	1.57%	1.32%	0.99%	1.32%	1.45%	1.27%	1.57%	1.30%	1.33%

doi:10.1371/journal.pone.0056517.t003

Table 4. Proportions of true toxicities covered by the first two predictions for training and test datasets of each test group

Test group	Training dataset	Test dataset
TG_1	65.52%	64.69%
TG_2	65.54%	64.52%
TG_3	65.43%	66.49%
TG_4	65.32%	65.83%
TG_5	65.48%	64.36%
TG_6	65.46%	65.71%
TG_7	65.55%	65.21%
TG_8	65.43%	65.82%
TG_9	65.61%	64.07%
TG_{10}	65.61%	64.79%

doi:10.1371/journal.pone.0056517.t004

Performance of the Method on the Test Dataset

For the 1,723 compounds in each test dataset $S_{te}^{(i)}(1 \leq i \leq 10)$, the toxicities of these compounds were predicted by the proposed method described in Section "Prediction method" based on the compounds in the training dataset $S_{tr}^{(i)}$. After processing by **Eq. 5**, seven prediction accuracies for each test dataset $S_{te}^{(i)}$ were obtained and were listed in the column with title $S_{te}^{(i)}$ of **Table 3**. It is observed that the 1st order prediction accuracies were all about 79.50%. Similar to the seven prediction accuracies for each training dataset $S_{tr}^{(i)}$, those of test dataset $S_{te}^{(i)}$ also followed a descending trend with the increase of the order number, implying that our method also arranged the candidate toxicities of samples in each test dataset quite well. According to **Eq. 7**, the average numbers of toxicities for the compounds in each test dataset were about 1.80. Thus, we still considered the first two predictions of each sample in $S_{te}^{(i)}(1 \leq i \leq 10)$ to calculate the proportions of true toxicities covered by these predictions, i.e., computing **Eq. 6** by taking $m=2$. Listed in column 3 of **Table 4** are ten proportions for ten test datasets, each yielding a probability of approximately 65%.

DISCUSSION

Understanding of the Toxicity Prediction Results

It is observed from **Table 3** that the performance of the method on ten test groups is similar. Thus, the first test group (i.e., TG$_1$) is used as an example to show how to interpret the toxicity predicting results in detail.

Our multiclass model achieved a quite promising performance using the chemical-chemical interactions data on test group TG$_1$ (see **Table 3** for details). For example, the compound 4-(N-methyl-N-nitrosamino)-1-(3-pyridyl)-1-butanone (CID000047289, NNK) shows positive results for five toxicity endpoints: T_1, T_2, T_3, T_5, and T_6. Our model accurately predicted these five kinds of endpoints, and provided the order predictions as $T_3 > T_2 > T_1 > T_6 > T_5 > T_4 > T_7$. The 7th label representing 'non-toxic' was ranked as the last, suggesting that this compound is very likely to have toxic effects. As stated in the Section "Chemical-chemical interactions", the interactive compounds derived from STITCH tend to have the same toxicity categories. 4-(Methylnitrosamino)-1-(3-pyridyl)-1-butanol (CID000104856, NNAL), an interactive compound of NNK, has toxicities T_2 and T_3, which are also shared by NNK. The alkyl N-nitroso group (see **Figure 2**) of these two compounds associates with the formation of DNA adducts, and induces lung cancer in laboratory animals [34], [35], [36]. Another example is trimethoprim (CID000005578), which is positive for five toxicity endpoints: T_1, T_2, T_4, T_5, and T_6. The prediction order of our model was $T_1 > T_6 > T_2 > T_5 > T_4 > T_3 > T_7$. This compound was considered to be a carcinogen according to chemical-chemical interactions, but the Accelrys Toxicity database [19] labeled this compound only as a mutagen. However, it is reasonable to assume this compound as a carcinogen because it has a genotoxic toxicophore-aromatic amine (see **Figure 2**) [5], [37], [38]. Typically, mutation is one of the first steps in the development of cancer [39].

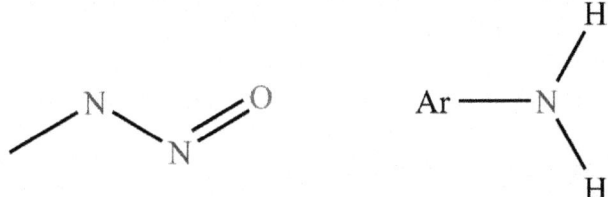

alkyl N-nitroso group   primary aromatic amine

Figure 2. The structures of the alkyl N-nitroso group and the primary aromatic amine group.

Tasosartan (CID000060919) is an angiotensin II (AngII) receptor blocker [40], which is labeled as a relatively "non-toxic" compound in the dataset. Using our model, the order prediction of this compound was $T_7 > T_1 > T_6 > T_2$. The 1st order prediction is "non-toxic", consistent with the experimental data available. Among seven interactive compounds in the training dataset retrieved from STITCH (see **Table 5**), the top five interactive compounds are "non-toxic", and their confidence scores are relatively high. However, the latter two interactive compounds are toxic, so tasosartan is predicted to have some toxicity effects in our model. However, the possibility of its possessing these toxicities is less than that of its not possessing toxicity (*i.e.*, "non-toxic").

Table 5. Details of Tasosartan's interactive compounds in the training dataset

Compound ID	Tag of toxicity class	Its interactive compound ID	Tag of toxicity class	Confidence score
CID000060919	T_7	CID000003749	T_7	679
CID000060919	T_7	CID000002541	T_7	670
CID000060919	T_7	CID000060921	T_7	669
CID000060919	T_7	CID000003961	T_7	667
CID000060919	T_7	CID000060846	T_7	658
CID000060919	T_7	CID000065999	T_1, T_6	643
CID000060919	T_7	CID000054738	T_1, T_2	172

doi:10.1371/journal.pone.0056517.t005

The predictions for NNK, trimethoprim, and tasosartan and the prediction accuracies of the method indicate that interactive compounds can share common toxicity with high probability, which assessment conforms to the results of predicting other attributes of compounds [15], [16]. The confidence scores of chemical-chemical interactions contribute significantly to the prediction of compound toxicity. As shown in **Table 5**, the interactive compounds of tasosartan with high confidence scores dominantly have the same toxicity as tasosartan. On the other hand, the predicted results for NNK, trimethoprim, and tasosartan reflect a limitation of our model: the judgment of "toxic" or "non-toxic" is based on a collective set of compounds with interactive information. However, some compounds with low confidence scores exist and they may contribute to the input of promiscuous interaction information to the final classification model. To address this issue, a future endeavor should introduce a threshold to the interaction confidence score and exclude "noisy" information to obtain a more accurate prediction.

Moreover, many more compounds are without chemical-chemical interactions in the original Accelrys Toxicity database. It is expected that the problem of predicting compound toxicity can be solved more favorably by the method as increasing amounts of chemical-chemical interaction information become available.

Analysis of the Relationship between Different Chemical Toxicity Effects

In the Accelrys Toxicity Database, there are 3,607 compounds with more than two types of toxicity effects and 3,475 compounds with exact two effects (refer to **Figure 1**). We analyzed the number of common compounds belonging to two categories, and the ratio of the number of common compounds to the number of non-overlapping compounds of the two categories (see **Table 6**). It can be found that the intersection of T_5 ("Reproductive Effects", cf. **Table 1**) and T_6("Multiple Dose Effects") is the largest, sharing 26.6% of common compounds. The overlapping compounds suggest that there may be a causal relationship between the two categories. Specifically, the reproductive effects may cause multiple dose effects, *i.e.*, reproductive toxicities may be cumulative, and hence be regarded as showing multiple dose effects in the meantime. The followed instances of correspondence between two categories are T_2("Mutagenicity") vs. T_3 ("Tumorigenicity") and T_1 ("Acute Toxicity") vs. T_6 ("Multiple Dose Effects"). Since, in many cases, mutation is one of the first steps in the development of cancer[39], we took T_2 ("Mutagenicity") vs. T_3 ("Tumorigenicity") as an example to study the relationship between the two toxic categories.

Table 6. The details of common compounds belonging to two categories

Tag of toxicity class	T_1	T_2	T_3	T_4	T_5	T_6
T_1	12,633[a]	3,483 (22.8%)[b]	1,485 (11.0%)	2,027 (15.6%)	2,075 (15.9%)	3,446 (25.7%)
T_2		6110	1,720 (25.7%)	1,213 (16.7%)	1,336 (18.4%)	1,723 (20.1%)
T_3			2293	570 (14.0%)	753 (18.6%)	781 (13.7%)
T_4				2353	731 (17.7%)	897 (15.9%)
T_5					2501	1,409 (26.6%)
T_6						4,198

[a]The number of common compounds belonging to two categories.
[b]The number in parenthesis means the ratio of the number of common compounds to the number of non-overlapping compounds of the two categories.
doi:10.1371/journal.pone.0056517.t006

From the viewpoint of mechanism of action, carcinogens can be classified into genotoxic or epigenetic carcinogens. Genotoxic carcinogens can bind covalently to DNA, and many known mutagens belong to this category. In the dataset, there are 1,720 common compounds with simultaneous toxicity T_2 ("Mutagenicity") and T_3 ("Tumorigenicity"). The Structural alerts (SAs) provided by Benigni [37], which are molecular functional groups associated with a specific toxicity end point [38], were used here to gain insights into the correspondence of the two toxic effects. As summarized in Table S1, we illustrated a few examples for each of the matched SAs.

As previously mentioned, not all of the mutagens are carcinogens. For example, α,β-unsaturated carbonyl compounds can interact with DNA by Michael addition, then lead to mutagenic and carcinogenic responses [37], *e.g.* acrylamide (CID000006579) and 2-butenal (CID000447466). However, if an α,β-unsaturated carbonyl compound has conformational constraints or alkyl groups at the site of nucleophilic attack, the compound would be prone to reaction via Schiff base formation [41]. This change may only generate the

DNA-adducts, but not undergo the following carcinogenic process [37]. This means that this kind of compound has no carcinogenicity, *e.g.* (E)-2-methyl-2-butenal (CID005321950) and 2-propylacrolein (CID000070609).

Epigenetic carcinogens do not usually bind directly to DNA, but have a large variety of different and specific mechanisms, and behave negatively in the standard mutagenicity assay [42]. Thus, some compounds that can match nongeneric SAs [37] are only carcinogens, not mutagens (see **Figure 3**).

Nongeneric SAs	Examples
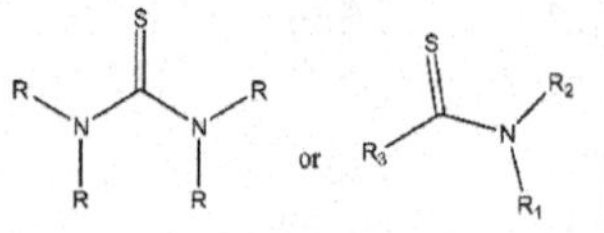 thiocarbonyl	diphenylthiohydantoin (CID000854150) rubeanic acid (CID002777982)
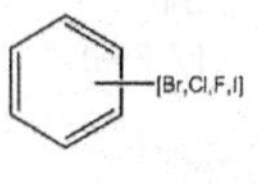 halogenated benzene	4-chlorothiophenol (CID000007815) 4-bromophenol (CID000007808)
Ar——[Br,Cl,F,I] halogenated PAH	4-hydroxy-2',4',6'-trichlorobiphenyl (CID000105036) 2,2',5,5'-tetrachlorobenzidine (CID000027465)
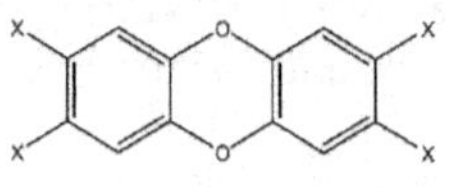 halogenated dibenzodioxins	1,2,3,4,6,7,8-heptachlorodibenzodioxin (CID000037270) 2,3,7,8-tetrabromodibenzo-4-dioxin (CID000039729)

Figure 3. Nongeneric SAs (Benigni) and some carcinogens matching these SAs.

CONCLUSIONS

In this study, a multi-classifier for six toxicity effects was built based on 17,233 compounds with their experimental toxicity information available and 323,432 pairs of mapped chemical-chemical interaction information extracted from the STITCH database. A new chemical entity can have multiple toxicity effects, so a multiclass toxicity prediction tool may prove to be practically more valuable to chemists than a traditional binary classification model. It can provide a better toxicity profile for a compound rather than merely indicating whether the compound has a specific toxic action or potential. The outstanding performance of our approach suggests that the multi-classification scheme is feasible and effective for *in silico* chemical toxicity prediction.

SUPPORTING INFORMATION

Table S1: SAs (Benigni) and examples matching SAs in our dataset

SAs	Examples
acyl halides	dimethylcarbamyl chloride (CID000006598) benzoyl chloride (CID000007412)
alkyl (C<5) or benzyl ester of sulphonic or phosphonic acid	trichlorfon (CID000005853) ethyl p-toluenesulfonate (CID000006638)
N-methylol derivatives	*N*-methylolacrylamide (CID000013543)
monohaloalkene	vinyl fluoride (CID000006339) 1-chloro-2-methylpropene (CID000010555)
S or N mustard	2,2',2"-trichlorotriethylamine (CID000005561) sulfur mustard (CID000010461)
propiolactones or propiosultones	propiolactone (CID000002365) 1,3-propane sultone (CID000014264)
epoxides and aziridines	3,4-epoxy-1-butene (CID000013586) aziridine (CID000009033)
aliphatic halogens	2-bromobutane (CID000006554) 2-iodobutane (CID000010559)
alkyl nitrite	isobutyl nitrite (CID000010958)

 α,β unsaturated carbonyls	acrylamide (CID000006579) 2-butenal (CID000447466)
 simple aldehyde	acetaldehyde (CID000000177) bromoacetaldehyde (CID000105131)
 quinones	phenylbenzoquinone (CID000009688) chloranil (CID000008371)
 hydrazine	1,1-dibutylhydrazine (CID000023902) hydrazine (CID000009321)
 aliphatic azo and azoxy	diazomethane (CID000009550) methylazoxymethanol acetate (CID005964719)
 isocyanate and isothiocyanate groups	toluene 2,4-diisocyanate (CID000011443) benzyl isothiocyanate (CID000002346)
 alkyl carbamate and thiocarbamate	ethyl butylcarbamate (CID000011577) pebulate (CID000014215)
polycyclic aromatic hydrocarbons	dibenzo(a,i)pyrene (CID000009106) dibenzo(a,h)pyrene (CID000009108)
heterocyclic polycyclic aromatic hydrocarbons	benzo(f)quinoline (CID000006796) benzo(b)naphtho(2,1-d)thiophene (CID000009198)
 alkyl and aryl N-nitroso groups	2-methylnitrosopiperidine (CID000023677) N-nitroso-N-butyl-N-propylamine (CID000032965)
 triazene groups	zidovudine (CID000035370)

isocyanate and isothiocyanate groups	toluene 2,4-diisocyanate (CID000011443) benzyl isothiocyanate (CID000002346)
alkyl carbamate and thiocarbamate	ethyl butylcarbamate (CID000011577) pebulate (CID000014215)
polycyclic aromatic hydrocarbons	dibenzo(a,i)pyrene (CID000009106) dibenzo(a,h)pyrene (CID000009108)
heterocyclic polycyclic aromatic hydrocarbons	benzo(f)quinoline (CID000006796) benzo(b)naphtho(2,1-d)thiophene (CID000009198)
alkyl and aryl N-nitroso groups	2-methylnitrosopiperidine (CID000023677) N-nitroso-N-butyl-N-propylamine (CID000032965)
triazene groups	zidovudine (CID000035370)

aliphatic N-nitro group	dimethylnitramine (CID000020120) N-nitrodiethylamine (CID000023505)
α,β unsaturated aliphatic alkoxy group	sterigmatocystin (CID005284457) aflatoxin G1 (CID000014421)
aromatic nitroso group	2-nitrosofluorene (CID000017271) 3,2'-dimethyl-4-nitrosobiphenyl (CID000051168)
nitro-aromatic	4,4'-dinitrobiphenyl (CID000015216) 4-nitrophenyl (CID000119211)
primary aromatic amine, hydroxyl amine and its derived esters or amine generating group	4-aminophenyl (CID0000181201) N-hydroxy-4-aminobiphenyl (CID000081261) N-acetoxy-4-acetylaminobiphenyl (CID000091584) toluene 2,4-diisocyanate (CID000011443) DADI (CID000007069)

 aromatic mono- and dialkylamine	*N,N,N',N'*-tetramethyl-4,4'-methylenedia niline (CID000007567) 4-dimethylaminostilbene (CID000640024)
 aromatic *N*-alkyl amine	*N,N'*-diacetylbenzidine (CID000011942) 4-acetylaminobiphenyl (CID000019998)
 coumarins and Furocoumarins	6-methylcoumarin (CID000007092) coumarin (CID000000323)

AUTHOR CONTRIBUTIONS

Conceived and designed the experiments: LC JZ MYZ YDC. Performed the experiments: LC JL KRF. Analyzed the data: JL JZ MYZ. Contributed reagents/materials/analysis tools: LC JL KRF MYZ YDC. Wrote the paper: LC JL KRF.

REFERENCES

1.   Dubach UC, Rosner B, Sturmer T (1991) An epidemiologic study of abuse of analgesic drugs. Effects of phenacetin and salicylate on mortality and cardiovascular morbidity (1968 to 1987). N Engl J Med 324: 155–160. doi: 10.1056/nejm199101173240304

2.   "AstraZeneca Decides to Withdraw Exanta" (2006) Available:http://www.astrazeneca.com/Media/Press-releases/Article/20060214-AstraZeneca-Decides-to-Withdraw-Exanta.Accessed 2012 Sep 2.

3.   Wang WB, Zhao YP, Cong L, Jing H, Liao Q, et al. (2011) Clinical characters of gastrointestinal lesions in intestinal Behcet's disease. Chin Med Sci J 26: 168–171. doi: 10.1016/s1001-9294(11)60043-6

4.   Zheng M, Liu Z, Xue C, Zhu W, Chen K, et al. (2006) Mutagenic probability estimation of chemical compounds by a novel molecular electrophilicity vector and support vector machine. Bioinformatics 22: 2099–2106. doi: 10.1093/bioinformatics/btl352

5.   Wang Y, Lu J, Wang F, Shen Q, Zheng M, et al. (2012) Estimation of carcinogenicity using molecular fragments tree. J Chem Inf Model 52: 1994–2003. doi: 10.1021/ci300266p

6. Crammer K, Singer Y (2001) On the algorithmic implementation of multiclass kernel-based vector machines. Journal of Machine Learning Research 2: 265–292.

7. Dietterich TG, Bakiri G (1995) Solving multiclass learning problems via error-correcting output codes. Journal of Artificial Intelligence Research 2: 263–286.

8. Sharan R, Ulitsky I, Shamir R (2007) Network-based prediction of protein function. Mol Syst Biol 3: 88. doi: 10.1038/msb4100129

9. Bogdanov P, Singh AK (2010) Molecular function prediction using neighborhood features. IEEE/ACM Trans Comput Biol Bioinform 7: 208–217. doi: 10.1109/tcbb.2009.81

10. Kourmpetis YA, van Dijk AD, Bink MC, van Ham RC, ter Braak CJ (2010) Bayesian Markov Random Field analysis for protein function prediction based on network data. PLoS One 5: e9293. doi: 10.1371/journal.pone.0009293

11. Ng KL, Ciou JS, Huang CH (2010) Prediction of protein functions based on function-function correlation relations. Comput Biol Med 40: 300–305. doi: 10.1016/j.compbiomed.2010.01.001

12. Hu L, Huang T, Liu XJ, Cai YD (2011) Predicting protein phenotypes based on protein-protein interaction network. PLoS One 6: e17668. doi: 10.1371/journal.pone.0017668

13. Hu L, Huang T, Shi X, Lu WC, Cai YD, et al. (2011) Predicting functions of proteins in mouse based on weighted protein-protein interaction network and protein hybrid properties. PLoS One 6: e14556. doi: 10.1371/journal.pone.0014556

14. Gao P, Wang QP, Chen L, Huang T (2012) Prediction of Human Genes Regulatory Functions Based on Protein protein Interaction Network. Protein and Peptide Letters 19: 910–916. doi: 10.2174/092986612802084528

15. Hu LL, Chen C, Huang T, Cai YD, Chou KC (2011) Predicting Biological Functions of Compounds Based on Chemical-Chemical Interactions. PLoS ONE 6: e29491. doi: 10.1371/journal.pone.0029491

16. Chen L, Zeng WM, Cai YD, Feng KY, Chou KC (2012) Predicting Anatomical Therapeutic Chemical (ATC) Classification of Drugs by Integrating Chemical-Chemical Interactions and Similarities. PLoS ONE 7: e35254. doi: 10.1371/journal.pone.0035254

17. Kuhn M, von Mering C, Campillos M, Jensen LJ, Bork P (2008) STITCH: interaction networks of chemicals and proteins. Nucleic Acids Res 36: D684–688. doi: 10.1093/nar/gkm795

18. Kuhn M, Szklarczyk D, Franceschini A, Campillos M, von Mering C, et al. (2010) STITCH 2: an interaction network database for small molecules and proteins. Nucleic Acids Res 38: D552–556. doi: 10.1093/nar/gkp937

19. Accelrys Toxicity Database 2011.4. Accelrys Software Inc.: San Diego, CA.

20. DrugBank. Available: http://www.drugbank.ca/downloads. Accessed 2012 Sep 2.

21. HMDB. Available: http://www.hmdb.ca/downloads. Accessed 2012 Sep 2.

22. Du P, Li T, Wang X (2011) Recent progress in predicting protein sub-subcellular locations. Expert Review of Proteomics 8: 391–404. doi: 10.1586/epr.11.20

23. Cai YD, Lu L, Chen L, He JF (2010) Predicting subcellular location of proteins using integrated-algorithm method. Molecular Diversity 14: 551–558. doi: 10.1007/s11030-009-9182-4

24. Shao X, Tian Y, Wu L, Wang Y, Jing L, et al. (2009) Predicting DNA-and RNA-binding proteins from sequences with kernel methods. Journal of Theoretical Biology 258: 289–293. doi: 10.1016/j.jtbi.2009.01.024

25. Zeng Y, Guo Y, Xiao R, Yang L, Yu L, et al. (2009) Using the augmented Chou's pseudo amino acid composition for predicting protein submitochondria locations based on auto covariance approach. Journal of Theoretical Biology 259: 366–372. doi: 10.1016/j.jtbi.2009.03.028

26. Chen L, Cai YD, Shi XH, Huang T (2012) Analysis of Metabolic Pathway Using Hybrid Properties. Protein and Peptide Letters 19: 99–107. doi: 10.2174/092986612798472857

27. Esmaeili M, Mohabatkar H, Mohsenzadeh S (2010) Using the concept of Chou's pseudo amino acid composition for risk type prediction of human papillomaviruses. Journal of theoretical biology 263: 203–209. doi: 10.1016/j.jtbi.2009.11.016

28. Georgiou D, Karakasidis T, Nieto J, Torres A (2009) Use of fuzzy clustering technique and matrices to classify amino acids and its impact to Chou's pseudo amino acid composition. Journal of theoretical biology 257: 17–26. doi: 10.1016/j.jtbi.2008.11.003

29. Li BQ, Hu LL, Chen L, Feng KY, Cai YD, et al. (2012) Prediction of Protein Domain with mRMR Feature Selection and Analysis. PLoS ONE 7: e39308. doi: 10.1371/journal.pone.0039308

30. Jin L, Fang W, Tang H (2003) Prediction of protein structural classes by a new measure of information discrepancy. Computational Biology and Chemistry 27: 373–380. doi: 10.1016/s1476-9271(02)00087-7

31. Ivanciuc O (2008) Weka machine learning for predicting the phospholipidosis inducing potential. Current Topics in Medicinal Chemistry 8: 1691–1709. doi: 10.2174/156802608786786589

32. Ravetti MG, Moscato P (2008) Identification of a 5-protein biomarker molecular signature for predicting Alzheimer's disease. PLoS ONE 3: e3111. doi: 10.1371/journal.pone.0003111

33. Sun XD, Huang RB (2006) Prediction of protein structural classes using support vector machines. Amino acids 30: 469–475. doi: 10.1007/s00726-005-0239-0

34. Yuan JM, Koh WP, Murphy SE, Fan Y, Wang R, et al. (2009) Urinary levels of tobacco-specific nitrosamine metabolites in relation to lung cancer development in two prospective cohorts of cigarette smokers. Cancer Res 69: 2990–2995. doi: 10.1158/0008-5472.can-08-4330

35. Kitiporn P, Jan-Phillip M, Marcus O, Fabian S, Victor S, et al. (2008) Machine learning based analyses on metabolic networks supports high-throughput knockout screens. BMC Systems Biology 2: 67. doi: 10.1186/1752-0509-2-67

36. Church TR, Anderson KE, Caporaso NE, Geisser MS, Le CT, et al. (2009) A prospectively measured serum biomarker for a tobacco-specific carcinogen and lung cancer in smokers. Cancer Epidemiol Biomarkers Prev 18: 260–266. doi: 10.1158/1055-9965.epi-08-0718

37. Benigni R, Bossa C (2011) Mechanisms of chemical carcinogenicity and mutagenicity: a review with implications for predictive toxicology. Chem Rev 111: 2507–2536. doi: 10.1021/cr100222q

38. Benigni R, Bossa C (2008) Structure alerts for carcinogenicity, and the Salmonella assay system: a novel insight through the chemical relational databases technology. Mutat Res 659: 248–261. doi: 10.1016/j.mrrev.2008.05.003

39. Arcos JC, Argus MF, editors (1995) Multifactor interaction network of carcinogenesis - a "tour guide". Boston: Birkhauser. 1–20 p.

40. DrugBank. Available: http://www.drugbank.ca/drugs/DB01349. Accessed 2012 Sep 12.

41. Patlewicz GY, Wright ZM, Basketter DA, Pease CK, Lepoittevin JP, et al. (2002) Structure-activity relationships for selected fragrance allergens. Contact Dermatitis 47: 219–226. doi: 10.1034/j.1600-0536.2002.470406.x

42. Woo YT (2003) Mechanisms of action of chemical carcinogens, and their role in structure–activity relationships (SAR) analysis and risk assessment. In: Benigni R, editor. Quantitative Structure–Activity Relationship (QSAR) Models of Mutagens and Carcinogens. Boca Raton: CRC Press. 41–80.

Chapter 12

A HYBRID PROCESS MONITORING AND FAULT DIAGNOSIS APPROACH FOR CHEMICAL PLANTS

Lijie Guo and Jianxin Kang

Hebei Key Laboratory of Applied Chemistry, College of Environmental and Chemical Engineering, Yanshan University, Qinhuangdao, Hebei 066004, China

ABSTRACT

Given their potentially enormous risk, process monitoring and fault diagnosis for chemical plants have recently been the focus of many studies. Based on hazard and operability (HAZOP) analysis, kernel principal component analysis (KPCA), wavelet neural network (WNN), and fault tree analysis (FTA), a hybrid process monitoring and fault diagnosis approach is proposed in this study. HAZOP analysis helps identify the fault modes and determine process variables monitored. The KPCA model is then constructed to reduce monitoring variable dimensionality. Meanwhile, the fault features of the monitoring variables are extracted, so then process monitoring can be performed with the squared prediction error (SPE) statistics of KPCA. Then, multiple WNN models are designed through the use of low-dimensional sample data preprocessed by KPCA as the training and test samples to detect the fault mode online. Finally, FTA approach is introduced to further locate the fault root causes of the fault mode. The proposed approach is applied to process monitoring and fault diagnosis in a depropanizer unit. Case study results indicate that this approach can be applicable to process monitoring and diagnosis in large-scale chemical plants. Accordingly, the approach can serve as an early and reliable basis for technicians' and operators' safety management decision-making.

INTRODUCTION

The chemical industry is one of the most important economic forces in world development [1]. The industry is adopting an increasingly large-scale, highly automated, and complex system because of increasing demands in terms of product quantity and production efficiency. However, serious consequences, such as major production losses, human injury, and environmental impact, can occur when errors emerge in chemical plants. Consequently, a significant amount of attention has been directed to the reliability and safety of the system in the chemical industry [2, 3].

A chemical process is broadly classified into a normal condition, abnormal condition, and fault condition. An abnormal condition is a range of abnormal operating states that are beyond the normal state but lack automated shutdowns [4]. This condition can occur in the system when the actual conditions deviate from original design conditions because of a slight fluctuation in variables or disturbance. If the abnormal condition is not monitored and handled in a timely manner, it could transform into a fault condition. Therefore, early monitoring and diagnosis of an abnormal condition are critical, so that appropriate actions can be taken to avoid fault.

In early times, process fault diagnosis completely relies on the domain knowledge of experts because of the lack of advanced monitoring devices and diagnosis approaches. As a result, faults cannot be monitored and diagnosed in a timely and accurate manner because of limitations in human ability. Over the past few years, many researchers have focused on process monitoring and fault diagnosis approaches to ensure process system safety [5–7]. Various techniques for process monitoring and fault diagnosis have been developed, such as mathematical-based models and knowledge-based and data-driven techniques. Mathematical-based approach was first proposed. However, the application of this approach is also limited because an accurate mathematical model is difficult to achieve or may even be unavailable for some complex industry plants. With advancements in computer control and artificial intelligence, data-driven and knowledge-based approaches have been developed in recent years. These approaches are constructed on the basis of the historical information of process variables and priori knowledge, respectively. Compared with the mathematical-based approach, accurate process models do not require to be established in the above two approaches. Therefore, data-driven and knowledge-based approaches are widely applied in process monitoring and fault diagnosis in industry plants [8].

An artificial neural network (ANN) is an extensively used knowledge-based approach in pattern recognition and classification of a nonlinear complex system because of its strong self-learning ability and nonlinear modeling. Among various ANN techniques, wavelet neural network (WNN) is a new class of neural network that has been used successfully in many studies [9]. Compared with other ANN techniques, WNN has a universal optimum capacity and a fast convergence speed. However, the fault diagnosis accuracy can decrease with the large architecture of WNN when the input data dimensionality is very large in the case of massive monitoring variables in large-scale chemical plants.

Various data-driven approaches have been used; examples are principal component analysis (PCA), independent component analysis (ICA), partial least squares (PLS), and techniques aided by subspace identification. Among these approaches, PCA is the most popular method applied in chemical process diagnosis. It is a multivariate statistical approach in which the fault feature of variables can be extracted and variable dimensionality can be reduced by analysis of the correlation among variables [10]. The state change of the system, that is, process monitoring, can then be monitored with PCA. Nevertheless, accurately identifying fault root causes with the conventional contribution plot approach in PCA is difficult because complicated process controls and recycle loops are common in industrial process [8].

In recent years, some researchers presented an improved PCA algorithm integrated with ANN for process monitoring and fault diagnosis. Chen and Liao proposed a process fault monitoring process based on a neural network and the PCA algorithm for chemical dynamic processes [11]. Kulkarni et al. developed a monitoring model of batch processes with a PCA-assisted generalized regression neural network [12]. Rusinov et al. built hierarchical neural networks integrated with PCA for fault diagnosis in chemical processes [13]. Jiang and Yan proposed a PCA model integrated with support vector data description for chemical process monitoring [14]. Although PCA and ANN have been successfully applied in process monitoring and fault diagnosis, PCA may not efficiently capture nonlinear features in the nonlinear process of the actual industrial complex system because PCA assumes that the relationship among variables is linear [15]. Kernel principal component analysis (KPCA) based on kernel function was first proposed by Schölkopf et al. to solve the problem caused by nonlinear data [16]. This approach maps the input space into a high-dimensional feature space and then computes the principal components (PCs) [17]. KPCA has proven more effectiveness than PCA in process monitoring and fault diagnosis [18].

Although the approaches discussed above have been employed in the fault diagnosis of some processes, in practice it is almost impossible for any method only to be successfully used for fault diagnosis of the large-scale chemical process system. For instance, it is extremely difficult to find the root causes of the fault by using alone PCA or KPCA. While the identification of fault root causes can be performed based on predefined knowledge base and previous experiences, in general, ANN or WNN used alone will lead to a large neural network size with long learning time and low diagnosis accuracy. Nonetheless, the combination of PCA and ANN also suffers from a drawback. There are usually a large number of the process variables in the complex chemical plants, so it is difficult to determine appropriate process monitoring variables for the specific fault mode by utilizing the above two approaches.

To address these problems, a hybrid approach based on hazard and operability (HAZOP) analysis, KPCA, WNN, and FTA for process monitoring and fault diagnosis is proposed in this study. HAZOP analysis is used as the first step to determine the fault mode and process variables that are monitored under fault condition. The KPCA model is constructed based on normal historical data obtained from process variables, and the squared prediction error (SPE) statistics is applied to process monitoring. Then, low-dimensional fault data preprocessed by KPCA are considered as the training and test samples of WNN. Finally, the FTA models are used as predefined knowledge base to further locate the fault causes of the fault modes. The proposed approach can be utilized to quickly monitor abnormal and fault conditions and effectively identify fault root causes. The information generated can then serve as a reliable decision-making basis for technicians and operators.

PROCESS MONITORING AND FAULT DIAGNOSIS APPROACH

The flowchart of the proposed process monitoring and fault diagnosis approach is shown in Figure 1. The herein proposed approach involves two steps:

   a)   establishment of the process monitoring and fault diagnosis model and

   b)   online application of the process monitoring and fault diagnosis model.

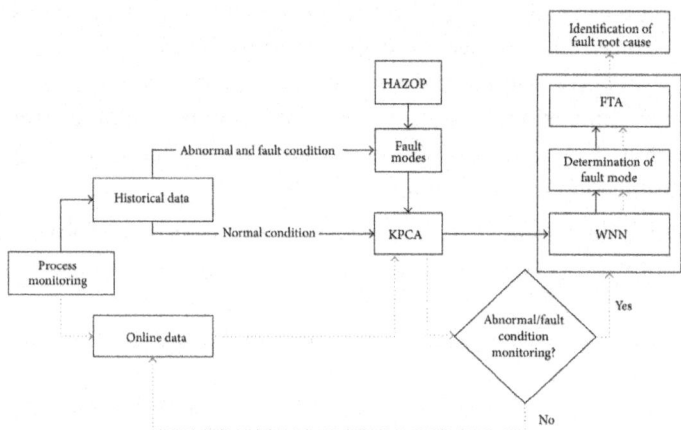

Figure 1: Flowchart of the process monitoring and fault diagnosis approach (black line: establishment procedure of the process monitoring and fault diagnosis model;

red dot line: online application of the process monitoring and fault diagnosis model; HAZOP: hazard and operability; KPCA: kernel principal component analysis; WNN: wavelet neural network; FTA: fault tree analysis).

Phase 1. Firstly, the fault modes and process variables monitored are determined by HAZOP analysis. Based on the historical monitoring data under normal condition, KPCA model is then constructed for reducing data dimensionality and obtaining SPE statistics. Next, multiple WNNs models are established to find fault modes through the use of fault historical data preprocessed by KPCA as the training and test samples. Finally, FTA method is introduced to identify fault root causes.

Phase 2. The preestablished KPCA model is applied to transform the high-dimensional online process monitoring data into lower dimensional data and calculate SPE statistics to monitor the abnormal and fault condition. Then, if SPE value exceeds the control limit, the data are fed to WNN models to pinpoint the fault mode. Moreover, the fault root causes by FTA can be further identified to effectively diagnose faults.

The procedure of the proposed approach is described as follows.

HAZOP

HAZOP analysis is currently recognized as the most widely used and preferred approach to identify hazards in the chemical process industry [19]. The plant studied is usually divided into some independent nodes to facilitate HAZOP analysis. For each node, some deviations are defined. These deviations consist of a guide word and a process variable, such as "higher temperature" and "lower temperature." Then, the causes and consequences of the potential hazards caused by these deviations are discussed by the HAZOP team. In this study, the deviations obtained from HAZOP analysis are used to build the fault modes that comprise the knowledge base in the fault diagnosis system.

Process Monitoring Based on KPCA

The basic idea of KPCA is to first map the input space into a highly dimensional feature space via nonlinear mapping and then compute the PCs on the feature space (x) [17]. This means that the data are performed PCA on the kernel feature space. Compared with that of the PCA approach, the main advantage of the KPCA approach is that it can extract more statistical features in the greatest degree relative to the original nonlinear data. In the present study, KPCA is used to reduce dimensionality and extract fault feature. Meanwhile, the SPE statistics is conducted to online monitor the incipient abnormal condition, and the control limit of SPE statistics under the normal condition is defined as Q_α.

When SPE value exceeds Q_α, the incipient abnormal condition can be detected and fault diagnosis can be performed by the following procedure. The detail KPCA procedure is presented in [15].

Fault Diagnosis Based on WNN

WNN is employed to quickly detect the fault modes of the abnormal condition. WNN is a new type of feedforward neural network that combines wavelet transform and ANN. The difference between WNN and conventional ANN is that a wavelet function is introduced to WNN as an activation function instead of the sigmoidal function. WNN integrates the advantages of both the wavelet multiscale time-frequency localization properties and self-learning of the neural network. Therefore, the convergence speed is faster and the universal approximation performance is stronger for WNN than for conventional ANN. Multiple fault modes exist in industrial processes. The diagnosis accuracy and convergence speed of WNN may thus be lower than those of multiple neural networks if a single WNN is constructed, which can cause a significantly large network topology. For this reason, multiple WNNs need to be separately constructed for different fault modes in this study.

A WNN with a three-layer network structure that consists of an input layer, a hidden layer, and an output layer was constructed and is shown in Figure 2.

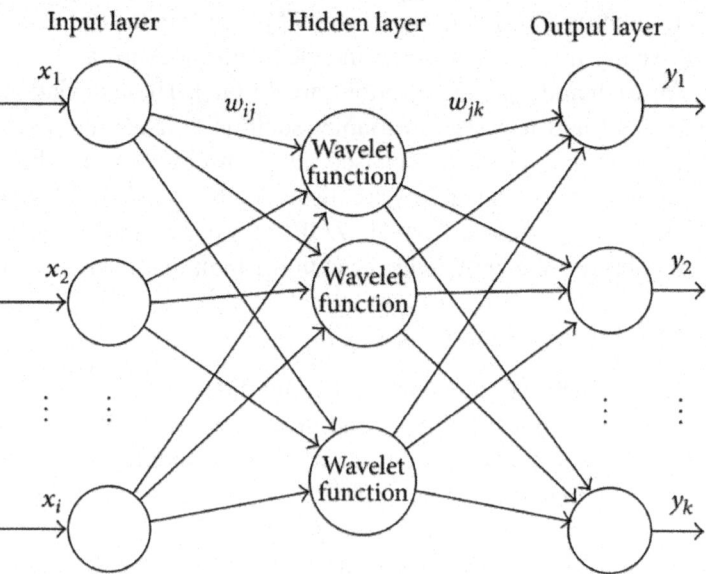

Figure 2: Network structure of the WNN.

After dimensionality reduction through KPCA, the fault historical process data were used as the input of the training and test samples in WNN. The code of the fault mode obtained by HAZOP analysis was considered as the output of WNN. In this case, the node numbers of the input layer and the output layer were determined with the PCs number of KPCA and the code digit of the fault modes, respectively. In this work, the numbered fault mode in ascending order is represented by a binary ASCII value, such as "0, 0, . . . , 1, 0, . . . , 0." The node number of the hidden layer can be calculated according to

$$s = \sqrt{(n + m)} + a, \qquad (1)$$

where s, n, and m denote the node number of hidden layers, input layers, and output layers, respectively; a is a specified parameter within the range $(0, 10)$. A Morlet wavelet function is selected as the activation function of the hidden layer; that is,

$$y = \cos{(1.75x)} e^{-x^2/2}. \qquad (2)$$

The outputs of the hidden layer $h(j)$ and the output layer $y(k)$ can be obtained by the following expressions:

$$h(j) = h_j \left[\frac{\sum_{i=1}^{n} w_{ij} x_i - b_j}{a_j} \right]; \quad j = 1, 2, \ldots, s,$$

$$y(k) = \sum_{j=1}^{s} w_{jk} h(j); \quad k = 1, 2, \ldots, m, \qquad (3)$$

where a_j and b_j are the factor of translation and dilation in wavelet function, respectively; w_{ij} is the weight parameters between the input and the hidden layer nodes; and w_{jk} is the weight parameters between the hidden and output layer nodes. The detailed descriptions of the above equations can be found in [20]. The historical data under abnormal and fault conditions are used to train the WNN structure. The preceding four parameters are updated and optimized continuously in the network training process until the absolute error value meets the desired goal. Finally, the ASCII value of the WNN predicted output is converted into an Arabic number that corresponds to the fault mode.

FTA

From the preceding procedure, the fault mode under the abnormal and fault condition can be detected with WNN. In many situations, however, locating the fault cause accurately is considerably difficult for technicians and operators because a fault mode usually relates to several fault causes. In this study, the

FTA approach is introduced to locate fault causes under abnormal and fault condition. FTA is a popular digraph approach used to perform quantitative risk assessment of a defined industrial process by combining the primary events with Boolean algebraic operators as indicated by the gates [21]. The output produced by WNN, that is, the fault mode, is regarded as a top event (TE) of FTA. Subsequently, the intermediate events (IEs) of the different levels are determined from top to bottom until all possible basic events (BEs) are identified by FTA. In this way, the fault tree models are constructed to use as predefined knowledge base. The fault root causes in a fault tree are often represented by the minimal cut sets of the fault tree [22]. Moreover, the higher is the probability of occurrence of the minimal cut set, the higher is also the probability that the event of the corresponding minimal cut set occurs. As a result, the probability of occurrence of each minimal cut set is evaluated quantitatively and ranked in descending order. In this case, the minimal cut set with the highest rank is considered as the most probable fault cause. Based on FTA results, technicians and operators can pinpoint the fault cause under the abnormal and fault condition to conduct appropriate preventive actions.

CASE STUDY

In this section, the proposed process monitoring and fault diagnosis approach was applied to a depropanizer unit. The schematic diagram of the depropanizer unit is illustrated in Figure 3. The liquefied petroleum gas (LPG) from the other unit enters the feed tank of the depropanizer and then pumped to the feeding preheater with a feeding pump, where it is heated to bubble point temperature. The LPG is fed into the 27th tray of the depropanizer. Both C2 and C3 fractions of the top of the depropanizer are condensed and then recycled to the reflux tank of the depropanizer. A part of the condensate is used as the reflux of the depropanizer. The remainder condensate is pumped to the deethanizer by the deethanizer feeding pump. C4 and C5 fractions from the depropanizer bottom are fed to the deisobutanizer.

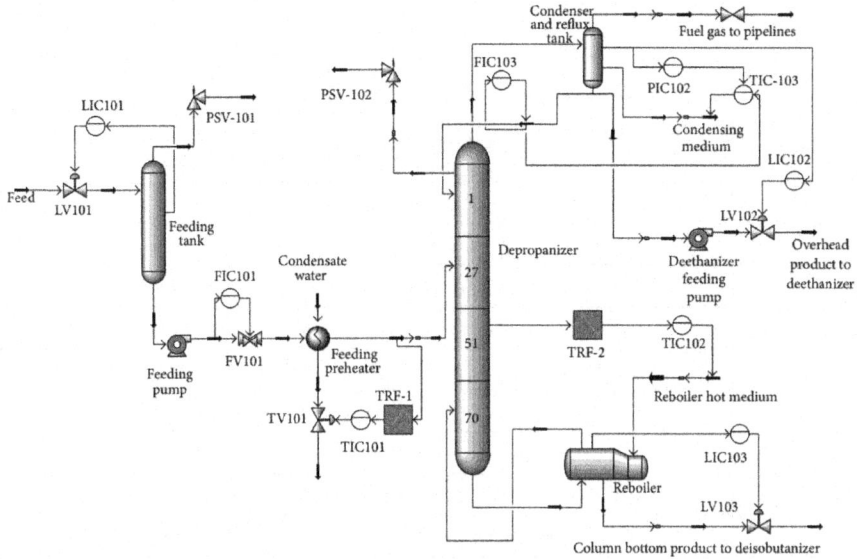

Figure 3: Schematic diagram of the depropanizer unit.

Data Sample Collection

From the HAZOP analysis results, 27 process monitoring variables (Table 1) were used, and the six types of deviations (Table 2) were selected as the fault modes for the case study. The dynamic process simulation model of the unit was established to simulate the normal operating condition, abnormal condition, and fault condition with UniSim software. After the simulation model ran for 50 minutes under normal operating condition, six types of disturbance signals were superimposed on the system to simulate the abnormal and fault conditions generated by fault modes. Fault samples were collected from the beginning of the disturbance. Each process variable under normal and fault conditions was sampled on a 20- and 4-second interval, respectively, in the simulation. A total of 150, 1125 samples were recorded under the normal operating condition, abnormal condition, and fault condition, respectively. A total of 1125 simulated sample data corresponding to each fault mode were randomly partitioned into a training set and a test set. The training set of each fault mode consists of 900 samples. The remaining 225 samples were considered as the test set. In addition, the sample data of any type of fault mode were regarded as an online validation set to verify the performance of the process monitoring and fault diagnosis model.

Table 1: Monitoring variable

Number	Variable
1	Feed flow of the unit/kmol h^{-1}
2	Pressure in the feeding tank/MPa
3	Liquid level in the feeding tank/%
4	Feed flow of the depropanizer/kmol h^{-1}
5	Feed temperature of the depropanizer/°C
6	Condensate water flow in the feeding preheater/kmol h^{-1}
7	Condensate water temperature in the feeding preheater/°C
8	Pressure on the top of the depropanizer/MPa
9	Temperature on the top of the depropanizer/°C
10	Overhead reflux flow of the depropanizer/kmol h^{-1}
11	Overhead reflux temperature of the depropanizer/°C
12	Liquid level of the overhead reflux tank/%
13	Pressure in the overhead condenser/MPa
14	Flow of the overhead product to the deethanizer/kmol h^{-1}
15	Cold fluid flow in the overhead condenser/kmol h^{-1}
16	C3 mole fraction in the overhead product/%
17	Sensitive tray temperature in the depropanizer/°C
18	Pressure at the bottom of the depropanizer/MPa
19	Temperature at the bottom of the depropanizer/°C
20	Liquid level at the bottom of the depropanizer/%
21	Temperature in the reboiler/°C
22	Flow rate from the reboiler back to the depropanizer/kmol h^{-1}
23	Fluid temperature from the reboiler back to the depropanizer/°C
24	Liquid level in the reboiler/%
25	Hot fluid flow in the reboiler/kmol h^{-1}
26	Flow of the column product to the deisobutanizer/kmol h^{-1}
27	C4 mole fraction in the bottom product/%

Table 2: Fault mode versus the output code in WNN

Number	Deviation/fault mode description	Binary ASCII code
1	Higher feed temperature of the depropanizer	0 0 0 0 0 1
2	No overhead reflux flow in the depropanizer	0 0 0 0 1 0
3	No flow of overhead product to the deethanizer	0 0 0 0 1 1
4	Lower temperature of sensitive tray in the depropanizer	0 0 0 1 0 0
5	Higher liquid level in the bottom of the depropanizer	0 0 0 1 0 1
6	Lower liquid level in the bottom of the depropanizer	0 0 0 1 1 0

Online Process Monitoring and Fault Diagnosis

The sets of sample data under normal condition were inputted into the KPCA model. The eigenvalue and cumulative contribution rate of the first seven PCs of the kernel matrix generated from KPCA are shown in Table 3. This table shows that the cumulative contribution rate (E value of 85%) of the first seven PCs is above 85%, so the number of the PCs was set as seven. That is, the number of dimensionalities of the sample data was reduced to 7 through KPCA from the original 27 dimensionalities. This finding indicates that dimensionality reduction of the nonlinear data through KPCA is obvious. Meanwhile, the SPE control limit (Q_α) was computed for process monitoring use.

Table 3: Cumulative contribution rate of the eigenvalue

Number	Eigenvalue	Cumulative contribution rate (%)
1	1.72	16.17
2	1.58	31.00
3	1.46	44.73
4	1.32	57.14
5	1.26	69.02
6	1.13	79.62
7	0.97	88.72

The online 2000 validation samples of fault mode 3 collected on a 3-second interval were inputted into the KPCA model to evaluate the performance of the online process monitoring model. The generated SPE (confidence limit of 99%) chart is shown in Figure 4. This figure shows that the SPE value at the 500th sample moment rapidly increases and exceeds the SPE control limit, and the system condition at the 904th sample moment becomes a fault condition. That is, the system condition at the 500th sample moment is a turning condition from the normal condition to the abnormal one. The results in the SPE chart agree with the predefined situation. In conclusion, the fault feature of the monitoring variables can be extracted effectively in KPCA monitoring.

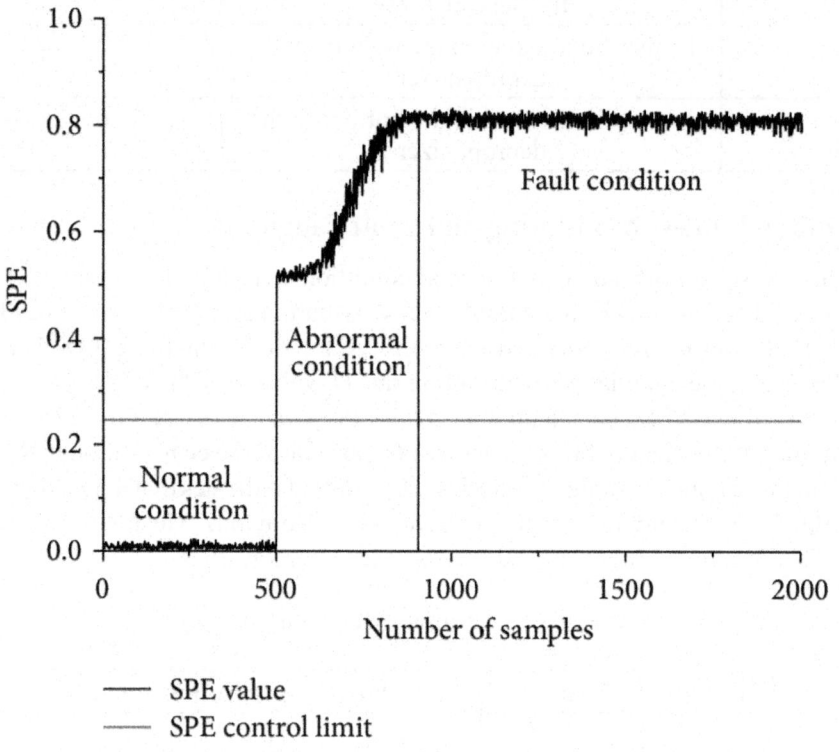

Figure 4: SPE chart for the online validation data set by KPCA.

Then, the causes resulting in abnormal and fault conditions were detected. Six types of fault modes were divided into two groups. In this way, each WNN corresponds to three types of fault modes. Two sets of 7-dimensional fault sample data preprocessed by KPCA were inputted into the WNN model. The fault mode code corresponding to the WNN output is represented by a six-digit ASCII code (Table 2). Therefore, the node number of the input layer and the output layer is 7 and 6, respectively. The node number of the hidden layer is determined as 14 according to (1). In this case, the WNN structure is 7-14-6. A total of 2700 training sets and 675 test data sets were used for WNN. The predicted absolute error of WNN is 0.05. Figure 5 shows the prediction results of the 900 validation samples with fault mode 3 in WNN. The Arabic numbers, from 1 to 6, represent the corresponding node sequence number of the WNN output layer in Figure 5. For example, data display the first node outputs obtained using the first WNN for validation samples in no. 1 subplot of Figure5(a). Figure 5 illustrates that the above 90% outputs of the first WNN obviously accord with the pattern of "0 0 0 0 1 1" within a specified limit of ±0.2. The outputs correspond with fault mode 3. However, the outputs of the second WNN are not followed by the regular pattern. The results of this case study show that the fault diagnosis accuracy of the proposed approach is very high.

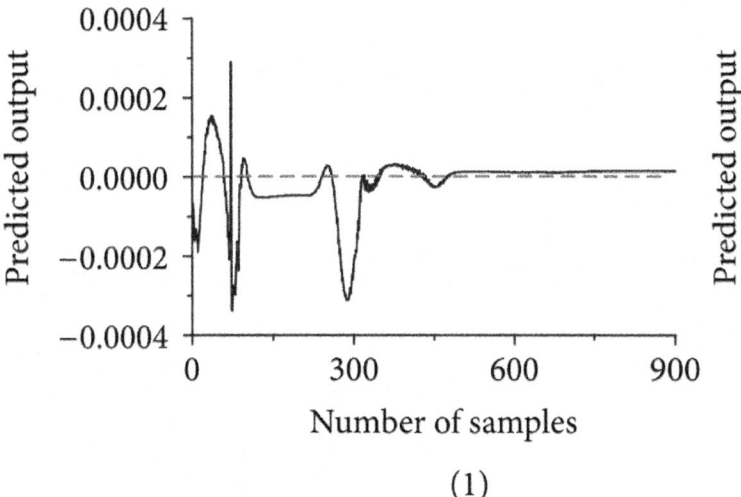

Number of samples

(1)

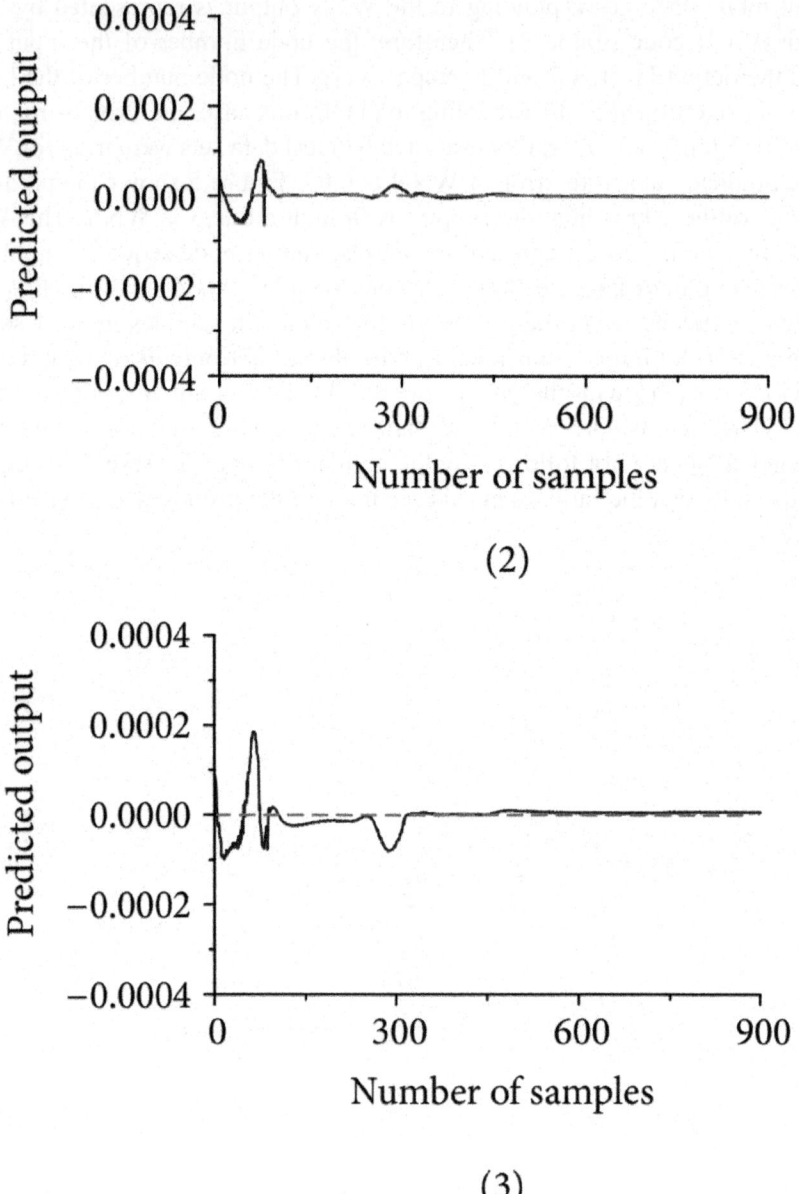

(2)

(3)

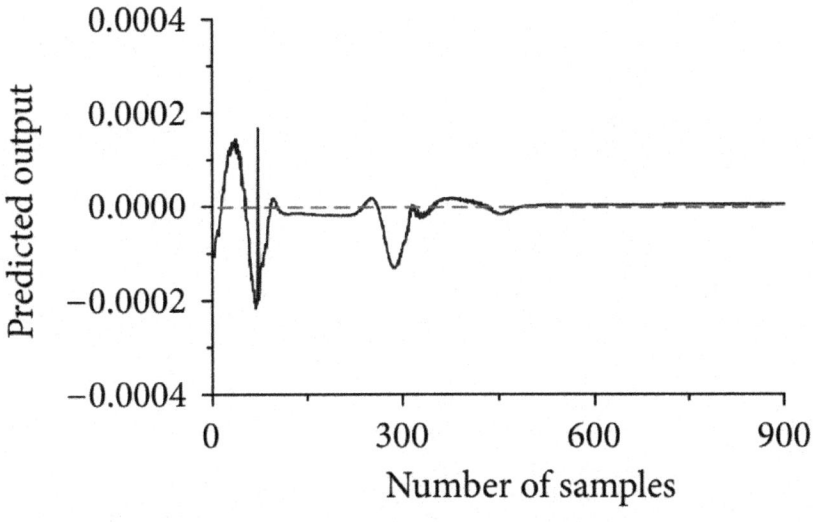

(4)

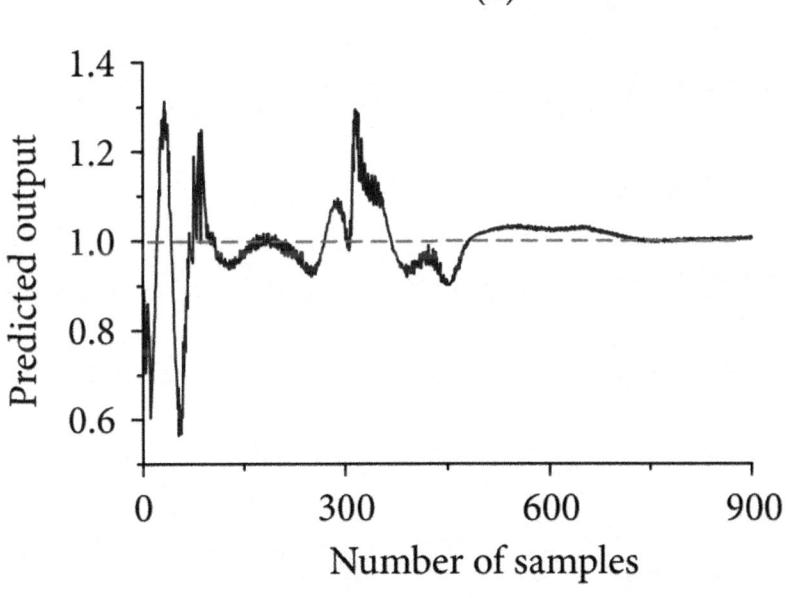

(5)

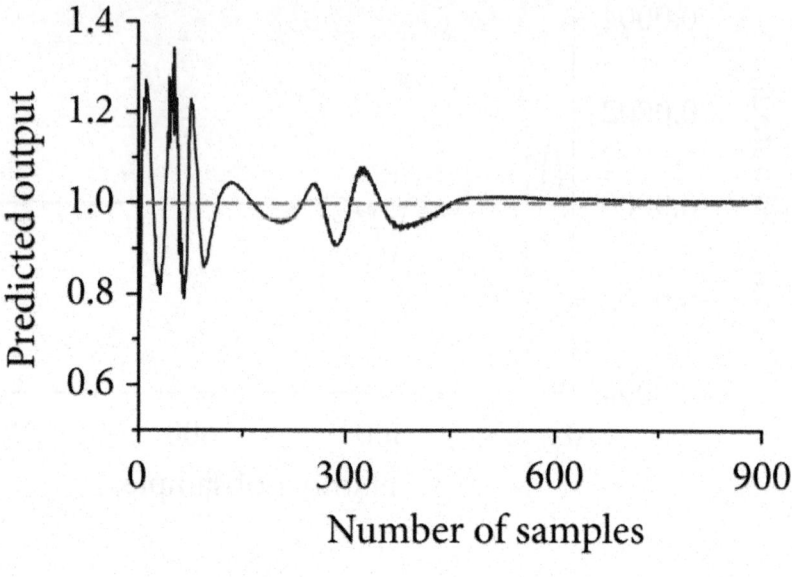

(6)

(a)

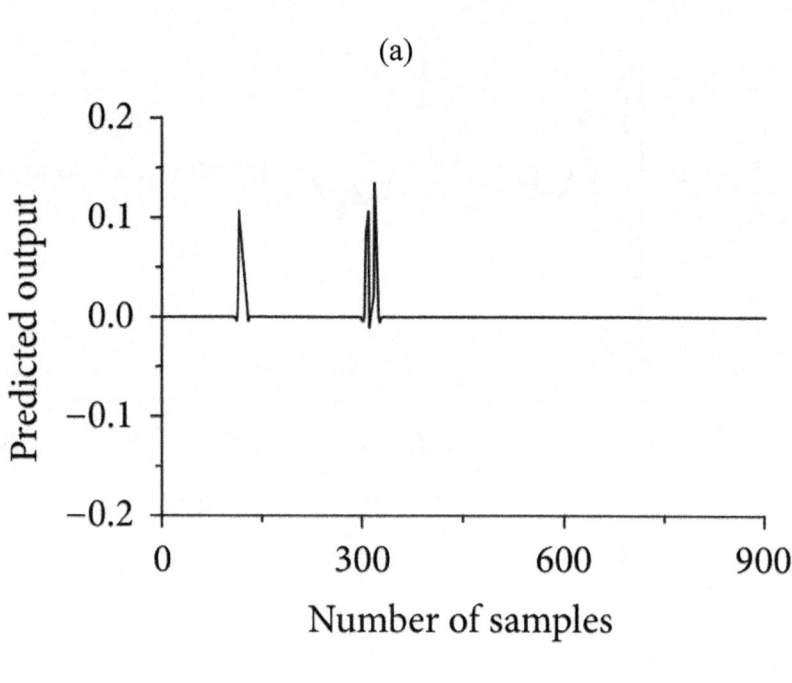

(1)

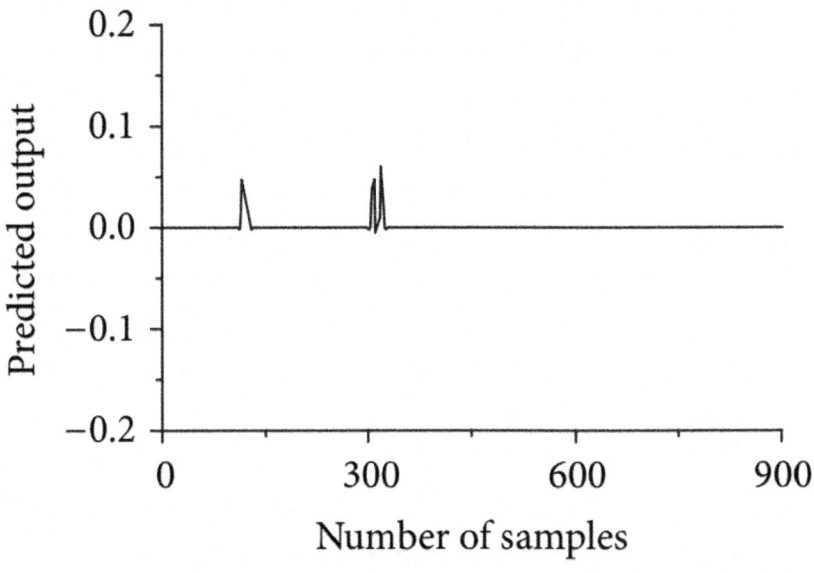

(2)

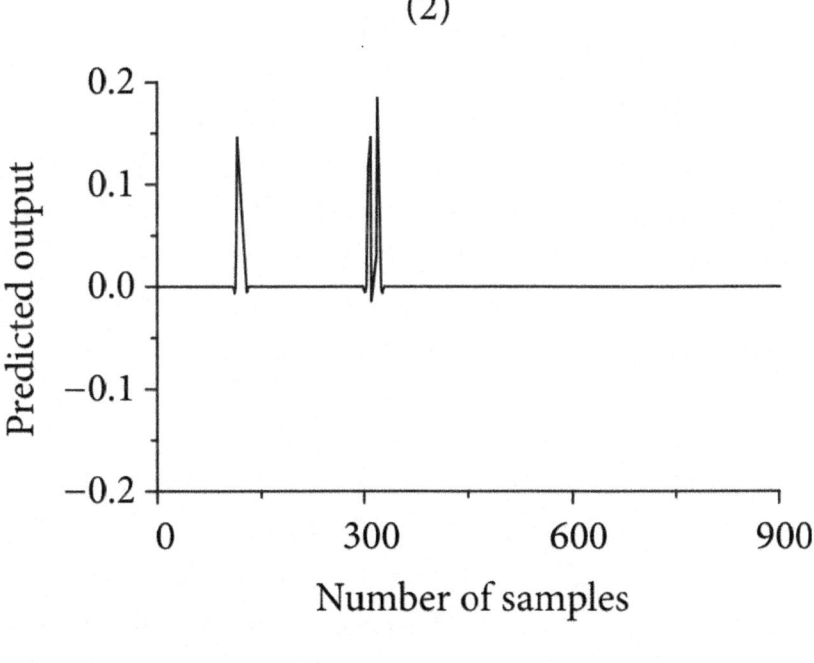

(3)

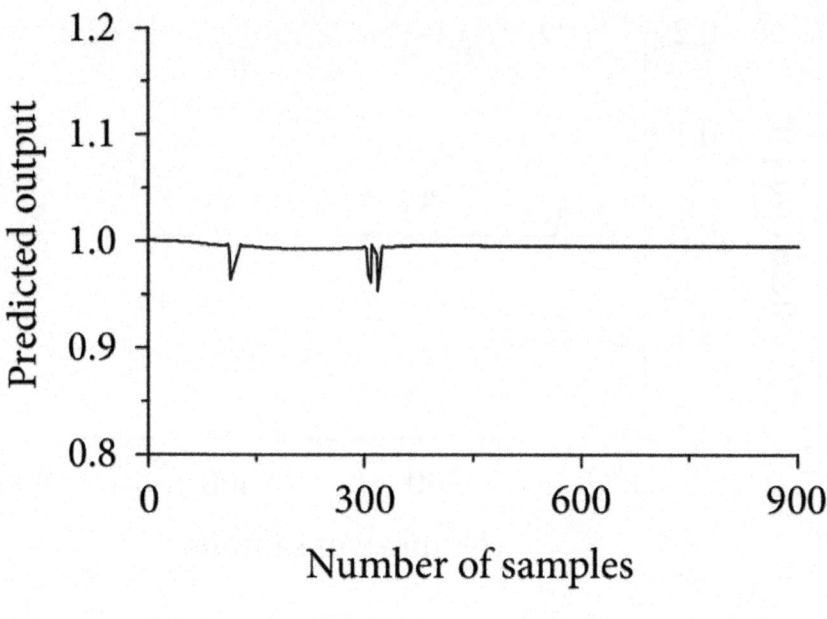

(4)

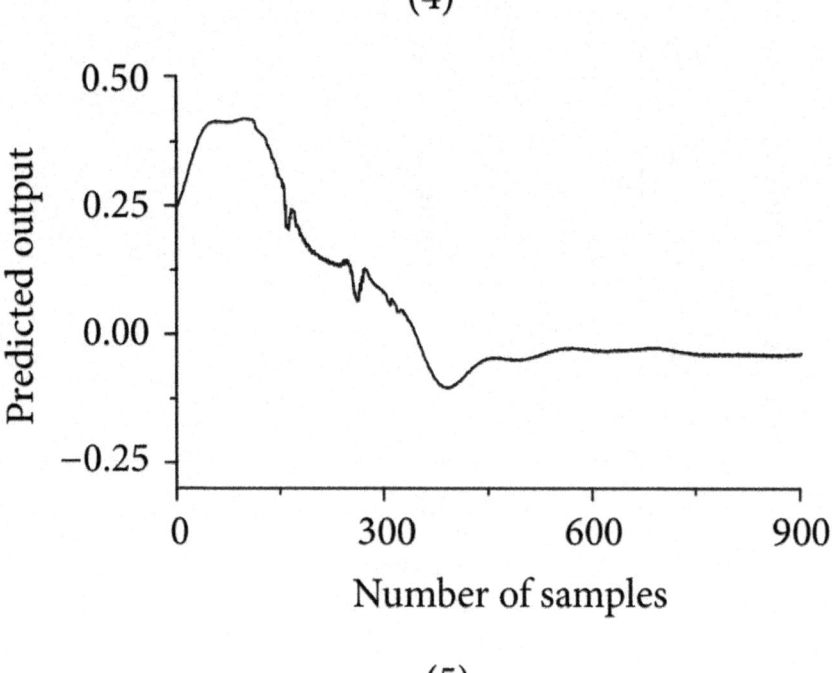

(5)

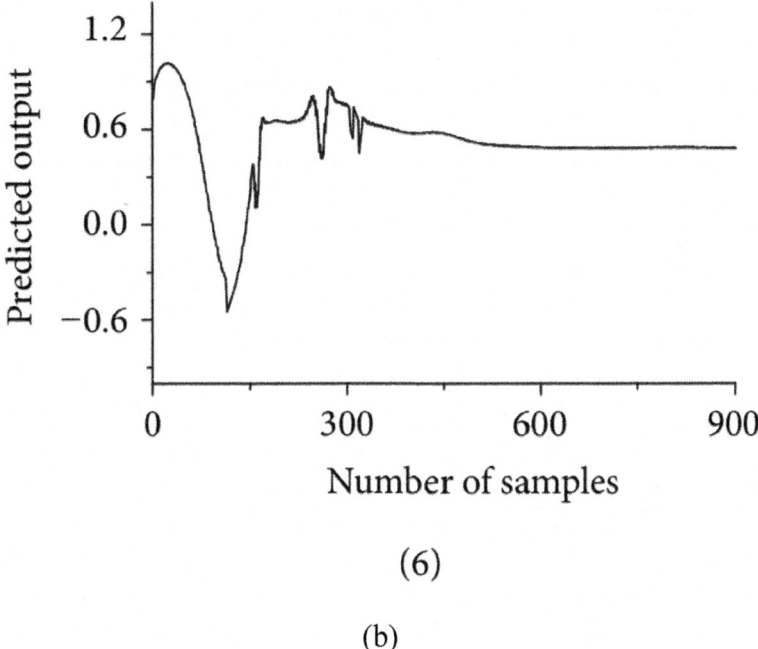

(6)

(b)

Figure 5: WNN outputs of the fault mode 3. (a) represents predicted outputs produced by the first WNN; (b) represents predicted outputs produced by the second WNN (black line: the predicted outputs of the WNN; red line: the desired outputs of the WNN).

After the fault mode under abnormal condition is identified, further diagnosis is required to pinpoint the root cause through FTA. A fault tree and each event related to fault mode 3 are illustrated in Figure 6 and Table 4, respectively. The fault mode 3, that is, no flow of the overhead product to the deethanizer, is defined as the TE of the fault tree. And then two IEs and five BEs are obtained by FTA. The minimal cut set of the fault tree and the occurrence probability (8000 h) corresponding to each minimal cut set are shown in Table 5. The table depicts that BE2 and BE1 are the prime consideration factors that result in the occurrence of TE. When the SPE value exceeds the control limit in the fault diagnosis system, based on the knowledge base of the FTA, technicians and operators can be assisted in effectively locating the root cause and then taking appropriate measures to eliminate the fault.

Table 4: Each event in the fault tree

Name	Event description
TE	No flow of the overhead product to the deethanizer

IE1	Pump (P-103) failure
IE2	Control loop (LIC 102) failure
BE2	Pump motor failure
BE1	Mechanical part failure in the pump
BE5	Liquid-level sensor failure
BE3	Controller (LIC 102) failure
BE4	Control value (LV 102) failure

Table 5: Probability of occurrence of the minimal cut set

Number	Minimal cut set	Probability of the occurrence (8000 h)
1	BE2	0.170258
2	BE1	0.158223
3	BE5	0.060869
4	BE3	0.043582
5	BE4	0.038749

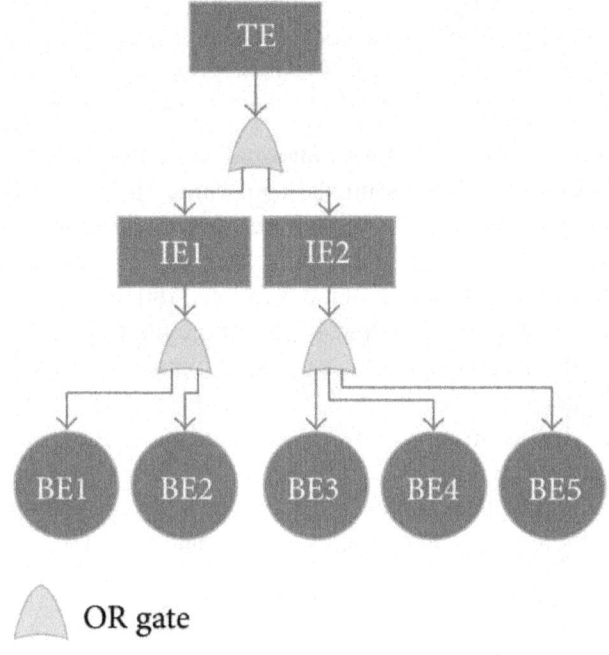

Figure 6: Fault tree with fault mode 3 (TE: top event; IE: intermediate event; BE: basic event).

CONCLUSIONS

In this study, a hybrid process monitoring and fault diagnosis approach is presented. To identify effectively potential abnormal and fault conditions, HAZOP analysis is used to analyze deviations as fault modes and process variables monitored. The KPCA method is developed to reduce data dimensionality and build the process monitoring statistics by extracting fault feature. According to HAZOP analysis, low-dimensional data corresponding to fault modes with the use of KPCA are regarded as the training and test samples of WNN. Then, the WNN model is constructed to detect the fault mode under an online abnormal and fault condition. To locate the root causes related to the fault mode, FTA is a particularly useful method to assist technicians and operators in quickly identifying potential risks. Process monitoring and fault diagnosis of a depropanizer unit are performed in a case study. The results show that the proposed hybrid approach is effective in ensuring process safety in large-scale chemical plants.

The sample dimensionality of the process monitoring variables in the KPCA model may be largely caused by process complexities in real plants. The dimensionality of the obtained kernel matrix may thus become significantly high; that is, the curse of dimensionality can occur. This problem in turn significantly affects the calculation speed of KPCA. Using precompression techniques for data, such as immune algorithm, can help address this problem. The number of monitoring variables can decrease significantly as a result.

On the other hand, if new fault modes or fault causes occur in process monitoring, the collected new data samples should be added to the WNN and fault tree model. When the same fault reoccurs, the system can detect the fault mode and pinpoint the fault causes.

ACKNOWLEDGMENT

The authors are thankful to the National Natural Science Foundation of China (Approved Grant no. 51205340) for providing financial support for this research.

REFERENCES

1. Q.-Q. Chen, Q. Jia, Z. W. Yuan, and L. Huang, "Environmental risk source management system for the petrochemical industry," Process Safety and Environmental Protection, vol. 92, no. 3, pp. 251–260, 2014.

2. L. J. Guo, J. J. Gao, J. F. Yang, and J. X. Kang, "Criticality evaluation of petrochemical equipment based on fuzzy comprehensive evaluation

and a BP neural network," Journal of Loss Prevention in the Process Industries, vol. 22, no. 4, pp. 469–476, 2009.

3. S. I. Ahmad, H. Hashim, and M. H. Hassim, "Numerical descriptive inherent safety technique (NuDIST) for inherent safety assessment in petrochemical industry," Process Safety and Environmental Protection, vol. 92, pp. 379–389, 2014.

4. C. Nan, F. Khan, and M. T. Iqbal, "Real-time fault diagnosis using knowledge-based expert system,"Process Safety and Environmental Protection, vol. 86, no. 1, pp. 55–71, 2008.

5. G. Wang, S. Yin, and O. Kaynak, "An LWPR-based data-driven fault detection approach for nonlinear process monitoring," IEEE Transactions on Industrial Informatics, vol. 10, no. 4, pp. 2016–2023, 2014.

6. S. Yin, S. X. Ding, X. C. Xie, and H. L. Luo, "A review on basic data-driven approaches for industrial process monitoring," IEEE Transactions on Industrial Electronics, vol. 61, no. 11, pp. 6414–6428, 2014.

7. S. X. Ding, "Data-driven design of monitoring and diagnosis systems for dynamic processes: a review of subspace technique based schemes and some recent results," Journal of Process Control, vol. 24, no. 2, pp. 431–449, 2014.

8. Y. W. Zhang, N. Yang, and S. P. Li, "Fault isolation of nonlinear processes based on fault directions and features," IEEE Transactions on Control Systems Technology, vol. 22, no. 4, pp. 1567–1572, 2014.

9. A. K. Alexandridis and A. D. Zapranis, "Wavelet neural networks: a practical guide," Neural Networks, vol. 42, pp. 1–27, 2013.

10. M. R. Maurya, R. Rengaswamy, and V. Venkatasubramanian, "Fault diagnosis by qualitative trend analysis of the principal components," Chemical Engineering Research and Design, vol. 83, no. 9, pp. 1122–1132, 2005.

11. J. H. Chen and C.-M. Liao, "Dynamic process fault monitoring based on neural network and PCA,"Journal of Process Control, vol. 12, no. 2, pp. 277–289, 2002.

12. S. G. Kulkarni, A. K. Chaudhary, S. Nandi, S. S. Tambe, and B. D. Kulkarni, "Modeling and monitoring of batch processes using principal component analysis (PCA) assisted generalized regression neural networks (GRNN)," Biochemical Engineering Journal, vol. 18, no. 3, pp. 193–210, 2004.

13. L. A. Rusinov, I. V. Rudakova, O. A. Remizova, and V. V. Kurkina, "Fault diagnosis in chemical processes with application of hierarchical neural

networks," Chemometrics and Intelligent Laboratory Systems, vol. 97, no. 1, pp. 98–103, 2009.

14. Q. C. Jiang and X. F. Yan, "Just-in-time reorganized PCA integrated with SVDD for chemical process monitoring," AIChE Journal, vol. 60, no. 3, pp. 949–965, 2014.

15. M. X. Jia, H. Y. Xu, X. F. Liu, and N. Wang, "The optimization of the kind and parameters of kernel function in KPCA for process monitoring," Computers and Chemical Engineering, vol. 46, pp. 94–104, 2012.

16. B. Schölkopf, A. Smola, and K.-R. Müller, "Nonlinear component analysis as a kernel eigenvalue problem," Neural Computation, vol. 10, no. 5, pp. 1299–1319, 1998.

17. J.-M. Lee, C. K. Yoo, S. W. Choi, P. A. Vanrolleghem, and I.-B. Lee, "Nonlinear process monitoring using kernel principal component analysis," Chemical Engineering Science, vol. 59, no. 1, pp. 223–234, 2004.

18. J. H. Li and P. L. Cui, "Improved kernel fisher discriminant analysis for fault diagnosis," Expert Systems with Applications, vol. 36, no. 2, pp. 1423–1432, 2009.

19. D. Ruiz, J. Cantón, J. M. Nougués, A. Espuña, and L. Puigjaner, "On-line fault diagnosis system support for reactive scheduling in multipurpose batch chemical plants," Computers & Chemical Engineering, vol. 25, no. 4–6, pp. 829–837, 2001.

20. X. C. Wang, F. Shi, L. Yu, and Y. Li, 43 Cases Analysis of MATLAB Neural Network, Beihang University Press, Beijing, China, 2014.

21. Z. Ruilin and I. S. Lowndes, "The application of a coupled artificial neural network and fault tree analysis model to predict coal and gas outbursts," International Journal of Coal Geology, vol. 84, no. 2, pp. 141–152, 2010.

22. Y. N. Papadopoulos, "Model-based system monitoring and diagnosis of failures using statecharts and fault trees," Reliability Engineering and System Safety, vol. 81, no. 3, pp. 325–341, 2003.

Chapter 13

ASSESSING SAFETY IN DISTILLATION COLUMN USING DYNAMIC SIMULATION AND FAILURE MODE AND EFFECT ANALYSIS (FMEA)

Suhendra Werner[1,2], Witt Fred[1] and Compart[1]

[1]Institute of Plant Design and Safety Technology, Technical University of Brandenburg LAS-BTU Cottbus, Haus 213, Burgerchausse 2-3, 03044-Cottbus, Germany

[2]Departement of Chemical Engineering, Faculty of Industrial Technology, Ahmad Dahlan University, 31. Prof DR. Soepomo, Janturan, Yogyakarta, DIY-Indonesia

ABSTRACT

Safety assessment becomes an important activity in chemical industries since the need to comply with general legal requirements in addition to meet safer plant and profit. This paper reviews some most frequently causes of distillation column malfunction. First, analysis of case histories will be discussed for providing guidelines in identifying potential trouble spots in distillation column. A dynamic simulation for operational failure is simulated as the basis for assessing the consequences. A case study will be used from a side stream distillation column to show the implementation of the concept. A framework for assessing safety in the column is proposed using Fault Mode and Effect Analysis (FMEA). Further, trouble-free operation in order to reduce the risk associated with column malfunction is described.

INTRODUCTION

Safety assessment becomes an important activity in industrial sustainability since the need to comply with general legal requirements. Also, an unsafe plant cannot be profitable over a period of time and due to losses of production as well as capital. Therefore, the objectives of a safety analysis are (CCPS, 2000).

- Reveal weaknesses of the plant
- Identify and describe relevant sequences of events
- Quantify frequencies of releases related to their consequence-potential
- Investigate safety gains from various possible system modifications and

- Improve the system if necessary (either alone or in combination with others).

   Accordingly, the framework for assessing safety is referred (Kister, 1997):

- Potential of flammable thermal of materials or mixtures under particular process.

- Trend of increasing temperature and/ or pressure at particular process and thermal production.

- Potentials of ignition and fire, either due to pressure, temperature, or concentration of processes.

- Toxic materials release.

   Therefore, this paper will describe a tool for safety assessment in distillation column. First, some troublesome distillation columns are discussed. The discussion is based on Kister's malfunction report histories (Jimoh, 2004). Then, one cause of possible malfunction is selected for the case study of the simulation of plant disturbance. Eventually, t the effect of malfunction is described.

THEORETICAL BACKGROUND

Safety in Distillation Column

Distillation column is the most commonly applied separation processes used in chemical industries. From safety point of view, there are significant numbers of safety disturbances in recent years based on Kister's surveys on column malfunction histories (Kister, 2003). The hazards in distillation emerge from high material contents and equipment complexity. Based on this fact, it is important to detect all important effects for safety and integrate into process model. The most important effects that must be investigated in distillation column are (Can, 2004):

- Influence of the hydrodynamic and mass transfer
- Control loop stability during nonstandard operation
- Effects of operational conditions on process safety
- Effectiveness of the protective systems

Dynamic Simulation for Safety Analysis in Distillation Column

In order to systematically characterize the effect of different operational disturbances, the use of dynamic modelling of the column can be a powerful tool for safety assessment taking into account that the malfunction is considered as reducing the optimum condition. Detailed dynamic simulation of operational failure (i.e., column malfunction) gives information concerning internal process behavior.

Therefore, the objectives of disturbance analysis are:

- Assessment of physical effects of disturbance in the plant
- Evaluation of possible malfunctions
- Technical know-how for relevant disturbances and associated effects, as well as risk on economy and emission
- Assessing alternatives for system optimization
- Assessing system protection

It is also the intention of further research in this field to attain the advantages on the dynamic simulation that the disturbance simulation can be integrated with the following objectives:

- Assessing the consequence
- Assessing the probability
- Measuring safety-related optimization for plant and processes

Risk Assessment

Risk is defined as a measure of human injury, environmental damage, or economic loss in terms of both the incident likelihood and the magnitude of the injury, damage, or loss. Risk analysis involves the development of an overall estimation of risk by gathering and integrating information about scenarios, frequencies and consequences and it is one major component of the whole risk management process of a particular enterprise. In the process of risk analysis, both qualitative and quantitative techniques can be used, as shown in Fig. 1. In practice, risk is often viewed as the product of the probability of an incident times consequence of the incident, as formulated in equation.

$$\text{Risk} = f\,(s,\,c,\,f) \qquad (1)$$

where s, c and f stand for scenario, consequence and frequency, respectively.

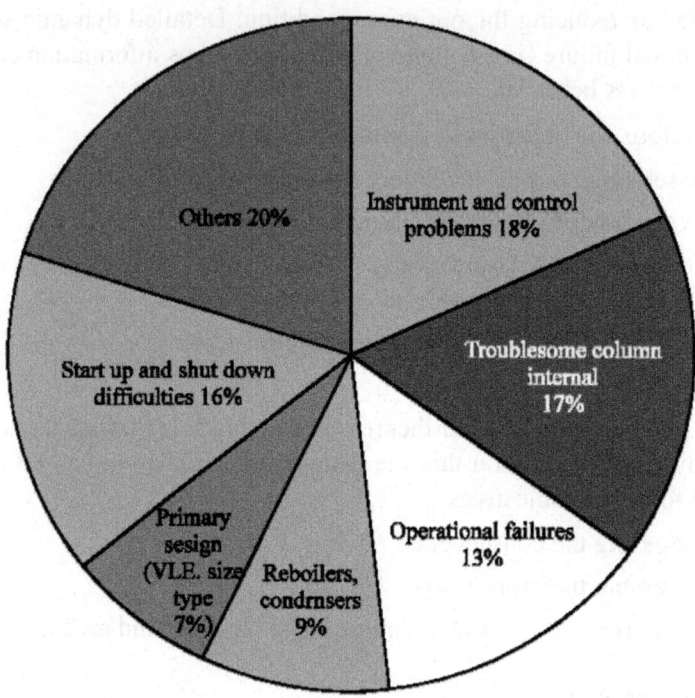

Figure. 1: A report on distillation malfunction histories (Jim, 2004)

A variety of techniques have been used for risk analysis in the chemical process industries including Safety Review, Checklist Analysis, Relative Ranking, What-if Analysis, Preliminary Hazard Analysis, Hazard and Operability Study (HAZOP), Failure Modes and Effects Analysis (FMEA), Fault Tree Analysis (FTA), Event Tree Analysis (ETA), Cause-Consequence Analysis (CCA), Human Reliability Analysis (HRA). Brief overviews of two methods will be discussed in the next paragraph, namely FMEA and ETA.

Failure Mode and Effects Analysis (FMEA)

FMEA is a systematic procedure in which each equipment failure mode is examined to determine its effects on the system and classify it according to severity and criticality. FMEA is an inductive method oriented toward equipment rather than process parameters. All of the failure modes for each item of equipment are tabulated with their effects, safeguards and related actions listed. An FMEA is especially useful to identify single failure modes that lead to an incident directly.

Event Tree Analysis (ETA)

An event tree is an inductive reasoning process that starts with an initiating event followed by the binary success or failure of subsequent safeguards, human responses and other safety measures to determine its possible outcomes. It is especially suitable to find possible outcomes of particular initial events and their respective probabilities with the data for initial events and subsequent protections and procedures.

Previous Literature

Elaahi and Luben (1983) has modelled the concept of distillation column with pressure relief system and performed some experimental work on pressure relief systems. The description of pilot plant measurement techniques to validate the simulation work has been described. Then, the simulation work was verified by an experimental work to validate the simulation of pressure relief as well as the case when the disturbance took place. He described also the scenarios that leads to the disturbances to occur and mechanism of pressure relief system. A dynamic simulation of operational failures were also performed by Can (2004) using Promps for methanol-water system.

Can et al. (2002) described a safety assessment method using Failure Mode and Effect Analysis (FMEA) that is applied in distillation column. These methods are started from the formulation of the event tree and fault tree. The Event Tree Analyses (ETA) starts from a defined initiating event and identifies potential consequences in a systematic way, where as the Fault Tree Analysis (FTA) starts from a defined undesirable (consequence) event and identifies basic events (like component malfunctions, operator errors etc.) which may lead to this undesirable event. This FMEA method requires raw data regarding explanation of the system/equipment function, fault-effect analysis, valuation of weak point, weak point elimination and minimization and risk potential minimization.

Case Study

The case study uses a column as a recycle part of nonseparated acetone from heavy ends column as shown in Fig. 2. Therefore, the separation task of this column is to separate valuable acetone into head column and the rest into the base column. The efficiency of separation is influenced by the available heat fed into column. In this case, live steam is fed from the base as the heat source.

The column feed stream is preheated in a heat exchanger by the hot base stream of the column and enter at tray 17. Live steam is injected into the base and the steam flow rate is temperature control at stage 27. The column has 35 valve trays. The head product is condensed in a heat exchanger with the vent passing through another heat exchanger to the atmosphere and the liquids are collected by reflux drum. The reflux is transferred under flow control back into the column on stage 35. The crude acetone is pumped through a heat exchanger for cooling under level control and on to acetone recovery plant. The base product passes the heat exchanger and runs on through another heat exchanger for further cooling and is then discharged into the in-plant effluent pit.

The column side stream is equipped in order to attain more middle boiling component, methanol and decrease its adverse effect in the base stream.

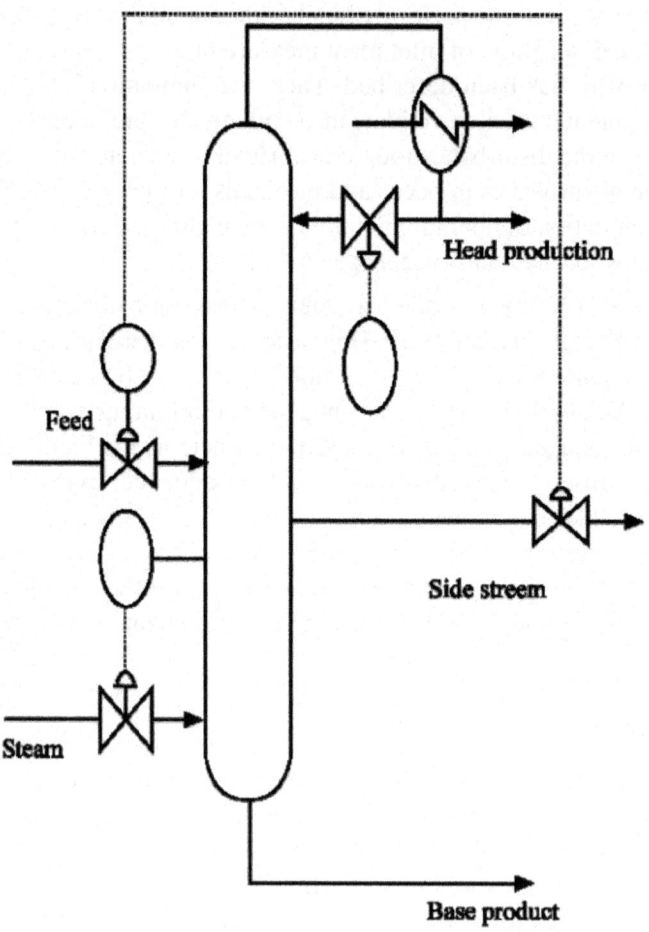

Figure. 2: Distillation column configuration for the case study

Table 1: Feed composition

Component	Composition (mass %)
Acetaldehyde	2.212
Acetic acid	1.752
Acetone	13.111
Diethyl ketone	0.098
Ethyl acetate	0.246
Ethanol	1.060
Formic acid	0.370
Methyl acetate	2.344
Methyl formate	3.770
Methyl ethyl ketone	0.565
Methanol	1.652
Propionic acid	0.136
Water	72.682

Such side stream configuration is also suggested consuming less energy (Alatiqi and Luyben, 1985; Elaahi and Luyben, 1983). The composition of feed stream is shown in the Table 1 and the coupled variables between process variable and output variables are shown in Table 2.

Proposed Safety Assessment

The assessment of risk potential refers to following framework:

- Consequence assessment with support of tabulated method and/or disturbance simulation.
- Consequence assessment with support of initiating events methods.
- The use of risk potential matrix for defining of not acceptable risks.
- The application of disturbance simulation for recognition of
- The optimal plant
- The optimized process
- The alternative/ renewed consequence of weak points
- Optimization of the plant and/ or process through technical as well as organization at framework.

The framework originates from the definition of initiating event, top event and the probability as well as the consequence. These are assessed and listed in a Table 2 and 3. The consequence addresses the intensity of side effect from particular system (e.g., effect to the environment). The general criteria of consequence are expressed qualitatively from very good implying that the improvement is not urgent to be applied, until bad condition of the system. The valuation and the description of consequence is shown in Table 1. Whereas, the valuation the probability is started from 1 (very low, probability = 10^{-7}/year) and ended with 9 (very high, probability = 10/Year), as shown in Table 4. Then, the combination of probability and consequence value results in the Risk Potential Index (RPI) which has significant meaning for the priority of improvement of a definite plant that can be plotted in RPI matrix.

Table 2: Coupled variables pairings for the case study

No	Controller	Process variable	Output variable
1	Pressure Control	Pressure on Stage 1	Cooling Water Flowrate
2	Condenser Level Controller	Liquid Level on Stage 1	Distillate Flow rate
3	Reflux/Head Controller	Temperature at Stage 4	Liquid Head Flowrate
4	Steam Controller	Temperature at Stage 20	Steam Flowrate
5	Sidestream Controller	Sidestream flowrate	Feed flowrate
6	Sump Level Controller	Sump Level	Bottom Flowrate

Table 3: Valuation of consequence

Consequence Value	Description
1-3	Good
4-5	Satisfied
6-9	Bad

Table 4: Valuation of probability

PV*	Frequency	Probability
1	10^{-7}/Year	Very low/ implausible
2	10^{-6}/Year	Low
3	10^{-5}/Year	
4	10^{-4}/Year	Moderate
5	10^{-3}/Year	
6	10^{-2}/Year	High
7	10^{-1}/Year	
8	1/Year	Very high
9	10/Year	

*PV: Probability Value

There are 3 regions in RPI matrix, acceptable regions, not acceptable region and acceptable region but with further evaluation or optimization. The development of RPI matrix will be shown later.

DISCUSSIONS

Dynamic Simulation Results

The dynamic simulation is performed using ASPEN Dynamic. The running time is 20 h. After 1 hour steady state operation, the cooling water is reduced to 10%. This disturbance leads to a pressure increase in the column. Reducing cooling water supply will cause a substantial reduction of the condensation rate. According to the condenser duty equation:

$$Q_{cond} = \dot{m} \bullet C_p \, \Delta\vartheta_c = k \bullet A \bullet A \, \Delta\vartheta_m \tag{2}$$

Therefore, at the constant heat supply to the column, reducing cooling medium will lead to increasing the temperature different. Then, an accumulation of vapor in the condenser will occur, causing pressure increase (Fig. 3).

Decreasing of cooling water leads to increasing pressure in the column, then temperatures of every stages will increase accordingly. The increased pressure and temperature in the column will lead to a partial condensation of the vapor phase at the constant of heat input. This leads to increasing temperature at head and base stages (Fig. 4). One possible consequence due to cooling water reduction is poor product quality (Fig. 5).

Assessing Safety

From safety point of view, increasing base level will give potential hazard due to increasing risk on malfunction.

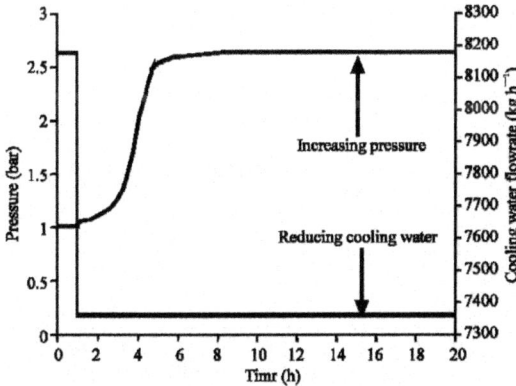

Figure. 3: Profile of increasing pressure after cooling water reduction.

It is true since most of Kister's malfunction histories emanate from bottom level failure due to the problems of excessive liquid raising.

According to Kister (2003) bottom problems cause 50% of the problems in distillation column. Therefore, potential liquid level rising above the base return inlet or bottom gas feed must be identified. Then, tower base level can be avoided. The potential consequence of problems on base tower are tower flooding, poor separation, instability and less vapor slugging through the liquid. All these problems can cause physical damage and threat to safety, as shown in Table 5. The values of semi-quantitative associated consequence are also given.

The prevention action for trouble free operation of distillation column in order to avoid those problems are reliable level monitoring, redundant system and good sump design (Kister, 1997). In addition, the issue on control assembly difficulties must be solved. The key success for this problems are a suitable control tray and pressure compensation for temperature control (Kister, 2003).

The event tree for cooling water reduction is developed as shown in (Fig. 6 and 7). And according to Table 4 and 5 above, the value of probability and consequence are determined. All values for possible risk are tabulated in the Fig. 8. A matrix of risk potential indices are created. Then the value of each risk potential index due to cooling water reduction can be defined. The values for all risk of consequence occurrences are 1, 7, 30, 28, 28, 30, 35 and 21, respectively. The darker the area in the matrix is, the higher risk will be. The information of potential risk is then documented in the FMEA data base. According to Fig. 6, all possible different operational failures will be included in FMEA data base.

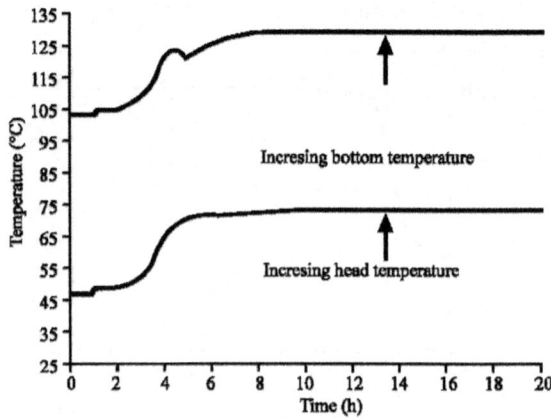

Figure. 4: Profile of increasing head and bottom temperatures after cooling water reduction.

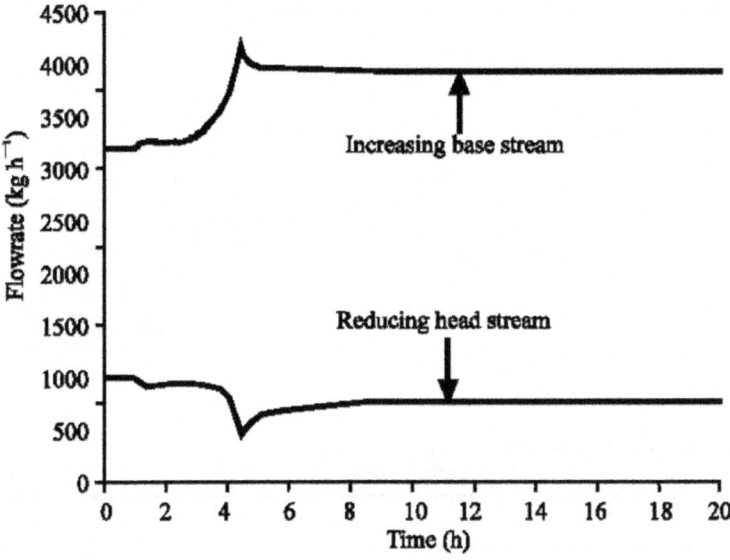

Figure. 5: Profile of increasing base stream and decreasing head stream after cooling water reduction.

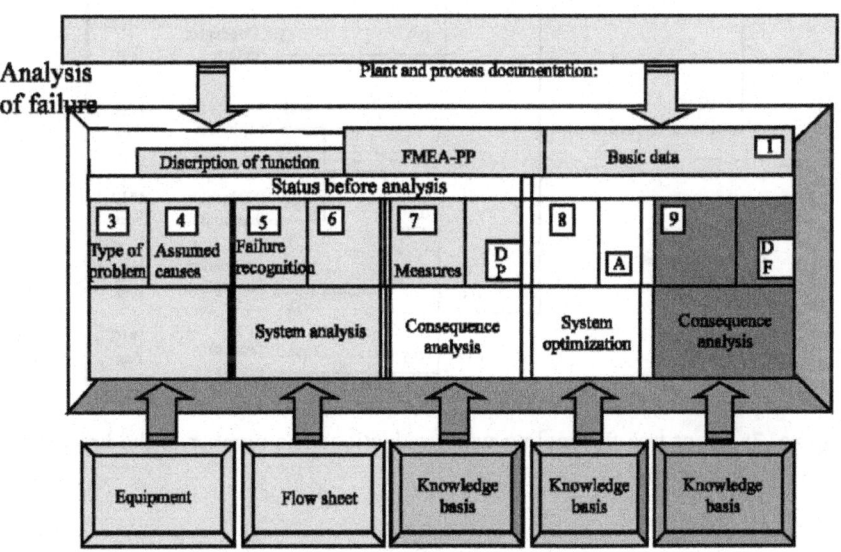

Figure. 6: FMEA data base for plant and process.

Table 5: The valuation of associated consequence for cooling water reduction

CV*	Type of Failure	Technical Example	Production Example
9	Safety and	Explosion	
8	environmental	Bursting of the column	
7	impact	Fire with destruction	
6		Fire (Pump/Heat exchanger/Boiler) with destruction	
		Emission through pressure release	
			Product interruption (long time)
5		Bottom failure (Flooding)	
	Mal function	Pumping failure	
		Reboiler failure	
		Material release due to leak	
			Product interruption
4		Emission through pressure relieve	
		Bottom Failure	
			Production interruption (short-time)
3	Function	Decreasing product quality (long-time)	
2	disturbance	Decreasing product quality	
1			Production interruption (short-time)

*CV: Consequence value

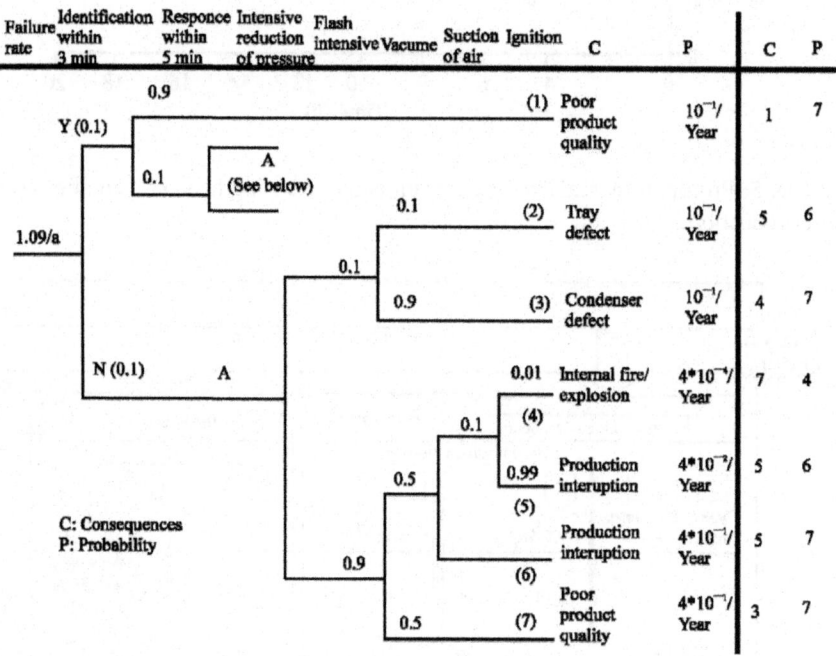

Figure. 7: Event tree for cooling water reduction. The number in the bracket represents the consequence.

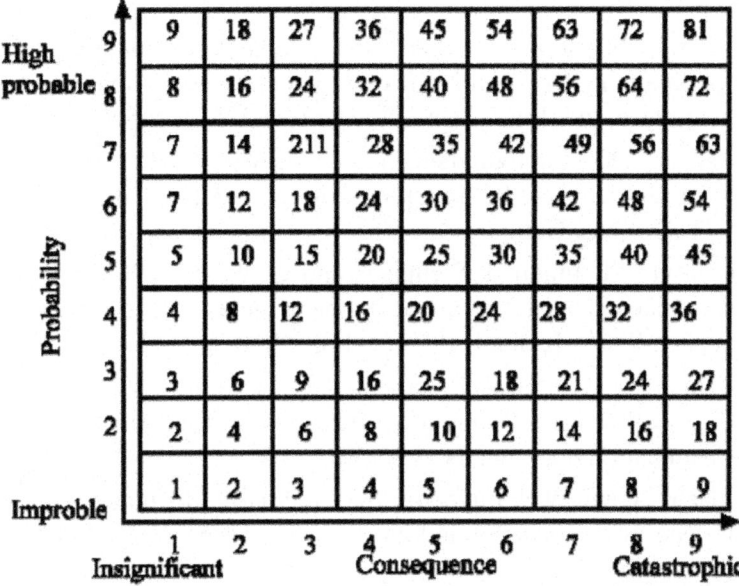

Figure. 8: A Matrix of Risk Potential Index.

With the aid of this data base, risks can be assessed for normal operational states as well as in case of operational failures.

CONCLUSIONS

The focus on recent assessment methods in distillation column is to identify the trends and to flag major regions of growing malfunction. The lessons from malfunction histories as well as simulation of column malfunction can save engineers and operators from failing into the same trap.

Using dynamic modeling of the column behavior during operational disturbances, the effect of such disturbances can be systematically characterized. The results could then be used for decision making in the development of new design regulations that would help to achieve hazard free operation or at least help in identifying the hazard potential of column under safety consideration.

Further combined analysis of the dynamic column behavior during non-standard operation together with the safety assessment method (such as FMEA) should give a deeper understanding of system safety. In addition, an assessment method for column safety should be integrated in an automatic way for thorough analysis of safety in distillation column.

REFERENCES

1.   Alatiqi, I.M. and W.L. Luyben, 1985. Alternative distillation configurations for separating ternary mixtures with small concentrations of intermediate in the feed. Ind. Eng. Chem. Process Des. Dev., 24: 506-507.

2.   Can, U., 2004. Zur Beherrschung sicherheitstechnisch relevanter Strungen beim Betrieb von Rektifikationskolonnen. Ph.D. Thesis, Dissertation, BTU-Cottbus, Germany.

3.   Can, U., M. Jimoh, J. Steinbach and G. Wozny, 2002. Simulation and experimental analysis of operational failures in a distillation column. Separa. Purifi. Technol., 29: 163-170.

4.   Center for Chemical Process Safety (CCPS), 2000. Guidelines for Chemical Process Quantitative Risk Assessment. American Institute of Chemical Engineering, New York.

5.   Elaahi, A. and W.N. Luyben, 1983. Alternative distillation configurations for energy conservation in four-component separations. Ind. Eng. Chem. Process Des. Dev., 22: 80-86.

6.   Jimoh, M., 2004. Entlastung von distillationkollonen im gestrten betrieb: Modllierung, simulation und experiment. Ph.D. Thesis, Dissertation, TU-Berlin, Germany.

7.   Kister, H.Z., 1997. Are column malfunctions becoming extinct-or will they persist in the 21st century?. Chem. Eng. Res. Des., 75: 563-589.

8.   Kister, H.Z., 2003. What caused tower malfunctions in the last 50 years?. Trans. IChemE., 81: 5-25.

CITATION

CHAPTER 1

Sun, K. , Bai, L. and Li, X. (2015) Analysis of the Chemical Safety Facility Investment Performance in China. Advances in Chemical Engineering and Science, 5, 102-109. doi: 10.4236/aces.2015.51011.

CHAPTER 2

Wang H, Zhang J, Lv G, Ma J, Ma P, Du G, et al. (2014) Preparation, Pharmacokinetics, Biodistribution, Antitumor Efficacy and Safety of Lx2-32c-Containing Liposome. PLoS ONE 9(12): e114688. doi:10.1371/journal.pone.0114688.

CHAPTER 3

El Haddad L, Ben Abdallah N, Plante P-L, Dumaresq J, Katsarava R, Labrie S, et al. (2014) Improving the Safety of Staphylococcus aureus Polyvalent Phages by Their Production on a Staphylococcus xylosus Strain. PLoS ONE 9(7): e102600. doi:10.1371/journal.pone.0102600.

CHAPTER 4

E. Labarthe, A. Bougrine, V. Pasquet and H. Delalu, "A New Strategy for the Preparation of N-Aminopiperidine Using Hydroxylamine-O-Sulfonic Acid: Synthesis, Kinetic Modelling, Phase Equilibria, Extraction and Processes,"Advances in Chemical Engineering and Science, Vol. 3 No. 2, 2013, pp. 157-163. doi: 10.4236/ aces.2013.32019.

CHAPTER 5

Juliano Missau, Amir J Scheid, Edson L Foletto, Sergio L Jahn, Marcio A Mazutti and Raquel C Kuhn, "Immobilization of commercial inulinase on alginate–chitosan beads," Sustainable Chemical Processes2014 2:13, DOI: 10.1186/2043-7129-2-13.

CHAPTER 6

Brian S. Flowers and Ryan L. Hartman, "Particle Handling Techniques in Microchemical Processes," Challenges 2012, 3, 194-211; doi:10.3390/challe3020194.

CHAPTER 7

Rajender S Varma, "Nano-catalysts with magnetic core: sustainable options for greener synthesis," Sustainable Chemical Processes2014 2:11, DOI: 10.1186/2043-7129-2-11.

CHAPTER 8

Susmit S Bapat, Clint P Aichele and Karen A High, "Development of a sustainable process for the production of polymer grade lactic acid," Sustainable Chemical Processes2014 2:3, DOI: 10.1186/2043-7129-2-3.

CHAPTER 9

Wei Wu1, Yuan-Tai Hsu2 and Po Chih Kuo1, "Scenario analysis of a bioethanol fueled hybrid power generation system," Sustainable Chemical Processes2014 2:5, DOI: 10.1186/2043-7129-2-5.

CHAPTER 10

Ramzan Naveed, Zeeshan Nawaz, Werner Witt and Shahid Naveed (2012). Systematic Framework for Multiobjective Optimization in Chemical Process Plant Design, Advances in Chemical Engineering, Dr Zeeshan Nawaz (Ed.), ISBN: 978-953-51-0392-9, InTech, DOI: 10.5772/36639.

CHAPTER 11

Chen L, Lu J, Zhang J, Feng K-R, Zheng M-Y, Cai Y-D (2013) Predicting Chemical Toxicity Effects Based on Chemical-Chemical Interactions. PLoS ONE 8(2): e56517. doi:10.1371/journal.pone.0056517.

CHAPTER 12

Lijie Guo and Jianxin Kang, "A Hybrid Process Monitoring and Fault Diagnosis Approach for Chemical Plants,"International Journal of Chemical Engineering, vol. 2015, Article ID 864782, 9 pages, 2015. doi:10.1155/2015/864782.

CHAPTER 13

Suhendra Werner, Witt Fred and Compart , 2007. Assessing Safety in Distillation Column Using Dynamic Simulation and Failure Mode and Effect Analysis (FMEA). Journal of Applied Sciences, 7: 2033-2039.

INDEX